PROCEEDINGS OF SYMPOSIA
IN PURE MATHEMATICS
Volume XXI

Representation Theory of Finite Groups and Related Topics

AMERICAN MATHEMATICAL SOCIETY
Providence, Rhode Island
1971

Proceedings of the Symposium in Pure Mathematics
of the American Mathematical Society

Held at the University of Wisconsin
Madison, Wisconsin
April 14–16, 1970

*Prepared by the American Mathematical Society
under National Science Foundation Grant GP-17009*

Edited by
IRVING REINER

International Standard Book Number 0-8218-1421-4
Library of Congress Catalog Number 79-165201
Copyright © 1971 by the American Mathematical Society
Printed in the United States of America

*All rights reserved except those granted to the United States Government
May not be reproduced in any form without permission of the publishers*

Contents

Preface	v
On the Degrees and Rationality of Certain Characters of Finite Chevalley Groups	1
By C. T. Benson and C. W. Curtis	
Types of Blocks of Representations of Finite Groups	7
By Richard Brauer	
Modular Representations of Some Finite Groups	13
By N. Burgoyne	
Some Connections Between Clifford Theory and the Theory of Vertices and Sources	19
By Edward Cline	
Finite Groups Admitting Almost Fixed-Point-Free Automorphisms	25
By Michael J. Collins	
Some Remarks on the Krull-Schmidt Theorem	29
By S. B. Conlon	
A Clifford Theory for Blocks	33
By E. C. Dade	
Jordan's Theorem for Solvable Groups	37
By Larry Dornhoff	
Operations in Representation Rings	39
By Andreas Dress	
Some Decomposable Sylow 2-Subgroups and a Nonsimplicity Condition	47
By Paul Fong	
Characters and Orthogonality in Frobenius Algebras	49
By T. V. Fossum	
The Number of Conjugacy Classes in a Finite Group	51
By P. X. Gallagher	
Sylow 2-Subgroups with non-Elementary Centers	53
By David M. Goldschmidt	
Axiomatic Representation Theory	57
By J. A. Green	
Real Representations of Split Metacyclic Groups	65
By Larry C. Grove	
On Some Doubly Transitive Groups	67
By Koichiro Harada	
Characterization of Rank 3 Permutation Groups by the Subdegrees	71
By D. G. Higman	

Symplectic Action and the Schur Index	73
BY I. M. ISAACS	
On Factorizable Groups	77
BY NOBORU ITO	
Lattices Over Orders	85
BY H. JACOBINSKI	
Faithful Representations of p Groups at Characteristic p	89
BY G. J. JANUSZ	
The Reflection Character of a Finite Group with a (B, N) Pair	91
BY ROBERT KILMOYER	
A Characterization of the Alternating Groups	95
BY TAKESHI KONDO	
Character Tables and the Schur Index	97
BY KARL KRONSTEIN	
Restriction of Representations Over Fields of Characteristic p	99
BY T. Y. LAM, I. REINER, AND D. WIGNER	
On the Suzuki and Conway Groups	107
BY J. H. LINDSEY II	
Matrix Questions and the Brauer-Thrall Conjectures on Algebras with an Infinite Number of Indecomposable Representations	111
BY L. A. NAZAROVA AND A. V. ROĬTER	
Group Rings of Infinite Groups	117
BY D. S. PASSMAN	
Characters of Finite Groups and Sets of Primes	123
BY WILLIAM F. REYNOLDS	
Bass-Order and the Number of Nonisomorphic Indecomposable Lattices Over Orders	127
BY KLAUS W. ROGGENKAMP	
The Modular Theory of Permutation Representations	137
BY L. SCOTT	
On the Affine Group Over a Finite Field	145
BY LOUIS SOLOMON	
Generalization of Green's Polynomials	149
BY T. A. SPRINGER	
A Splitting Principle in Algebraic K-Theory	155
BY RICHARD G. SWAN	
Classification of Simple Groups of Order $p \cdot 3^a \cdot 2^b$, p a Prime	161
BY DAVID WALES	
Direct Summands in Representation Algebras	165
BY W. D. WALLIS	
Representations of Chevalley Groups in Characteristic p	169
BY W. J. WONG	
Nilpotent Elements in Representation Rings	173
BY JANICE ROSE ZEMANEK	
Index	175

Preface

The symposium on Representation Theory of Finite Groups and Related Topics was held in Madison, Wisconsin, on April 14–16, 1970, in conjunction with a sectional meeting of the American Mathematical Society. The symposium was held in honor of Professor Richard Brauer, whose fundamental work in representation theory lies at the heart of most of the further developments in this topic.

These proceedings contain articles by the participants, based on their symposium lectures. The articles range from brief surveys of results to detailed outlines of proofs, and are intended to indicate the scope of current research in representation theory.

The organizing committee consisted of Professors Richard Brauer, Charles W. Curtis, Walter Feit, James A. Green, and Irving Reiner (chairman). The committee wishes to express its thanks to the National Science Foundation for its financial support of the symposium. We are also grateful to our colleagues at the University of Wisconsin for making us welcome in Madison. Finally, we thank the American Mathematical Society staff for helping with the arrangements of the symposium.

IRVING REINER

On the Degrees and Rationality of Certain Characters of Finite Chevalley Groups

C. T. Benson and C. W. Curtis

The results to be described concern the irreducible complex characters χ such that $\chi \in 1_B^G$, where B is a Borel subgroup of a finite Chevalley group or a twisted type. It is conjectured that these characters are rational on the whole Chevalley group, and that their degrees are polynomials in q, where q is the order of the finite field associated with the Chevalley group. Methods for proving these conjectures will be described, and affirmative answers to the conjectures will be stated for certain types of Chevalley groups.

In order to discuss the results from a uniform point of view, we axiomatize the situations as follows (see [4]).

DEFINITION. Let (W,R) be a finite Coxeter system, with finite Coxeter group W and set of distinguished generators R. A *system $\mathscr{S}$ of (B,N)-pairs of type (W,R)* consists of the Coxeter system (W,R), an infinite set of prime powers $\{q\}$, called *characteristic powers*, a set of positive integers $\{c_r\}_{r \in R}$, and for each characteristic power q, a finite group $G = G(q)$ with a (B,N)-pair such that

(i) $c_r = c_s$, for $r,s \in R$, if r and s are conjugate in W;
(ii) for each group $G = G(q) \in \mathscr{S}$, the index parameters ind $r = [B : B \cap B^r] = q^{c_r}$, $r \in R$;
(iii) if (W,R) is of type G_2, $c_r + c_s$ is even, $r,s \in R$.

To such a system corresponds a *generic ring* $A = A(u)$, which is an algebra with a free basis $\{a_w\}_{w \in W}$ over the polynomial ring $\mathfrak{o} = Q[u]$, satisfying, for $w \in W, r \in R$:

$$a_w a_r = a_{wr}, \quad l(wr) > l(w),$$

$$a_w a_r = u^{c_r} a_{wr} + (u^{c_r} - 1) a_w, \quad l(wr) < l(w).$$

AMS 1970 subject classifications. Primary 20C15; Secondary 20C05, 16A26.

For each homomorphism $f: \mathfrak{o} \to Q$, we can define the specialized algebra over Q, $A_f = Q \otimes_\mathfrak{o} A$, which has as a Q-basis the elements $\{a_{wf} = 1 \otimes a_w\}_{w \in W}$.

If f is the specialization $u \to q$ for a characteristic power q, then $A(q) = A_f \cong H_Q(G,B)$, the Hecke algebra (or centralizer ring) $eQGe \cong \text{Hom}_{QG}(1_B^G, 1_B^G)$, where $G = G(q) \in \mathcal{S}$, and $e = (1/|B|)\Sigma_{x \in B} x$.

For the specialization $u \to 1$, we have $A(1) \cong QW$.

Let $K = Q(u)$, $\bar{K}, \bar{Q}$ algebraic closures of K and Q respectively, and $\mathfrak{o}^*$ the integral closure of $\mathfrak{o}$ in $\bar{K}$. Using the proof of the deformation theorem of Tits, one can prove the following result:

LEMMA 1. *Let χ be an irreducible character of the generic algebra $A^{\bar{K}}$. Then $\chi(a_w) \in \mathfrak{o}^*$, for all $w \in W$. Let $f: u \to q$ or 1 define a specialization belonging to a characteristic power q or 1. Let $f^*: \mathfrak{o}^* \to \bar{Q}$ be an extension of f. Then the map $\chi_{f^*}: A_f^{\bar{Q}} \to \bar{Q}$ defined by*

$$\chi_{f^*}(a_{wf}) = f^*(\chi(a_w))$$

is an irreducible character of $A_f^{\bar{Q}}$. For a fixed extension f^* of f, the map $\chi \to \chi_{f^*}$ is a bijection between the irreducible characters of $A^{\bar{K}}$ and those of $A_f^{\bar{Q}}$.

From [3] we have:

PROPOSITION 1. *With the notations as in Lemma 1, each specialized character χ_{f^*}, viewed as a character of $H_{\bar{Q}}(G,B)$, is the restriction to $H_{\bar{Q}}(G,B)$ of a unique absolutely irreducible character ζ_{χ,f^*} of $G(q)$ (or W) such that $\zeta_{\chi,f^*} \in 1_B^{G(q)}$. Every irreducible constituent of 1_B^G is obtained in this way. The degree of ζ_{χ,f^*} is given by*

$$\deg \zeta_{\chi,f^*} = \frac{[G:B] \deg \chi}{\sum_{w \in W} \frac{1}{\text{ind } w} \chi_{f^*}(a_{wf}) \chi_{f^*}(\hat{a}_{wf})}$$

where $\hat{a}_{wf}$ is the basis element corresponding to w^{-1}, and $\text{ind } w = [B: B \cap B^w]$, $w \in W$. For the specialization $f_0: u \to 1$, the formula becomes $\deg \zeta_{\chi,f_0^*} = \deg \chi$.

DEFINITION. With the notations as above, let

$$d_\chi = \left(\sum_{w \in W} v(a_w)\right) \deg \chi \bigg/ \sum_{w \in W} \frac{1}{v(a_w)} \chi(a_w)\chi(\hat{a}_w),$$

where $v: A \to \mathfrak{o}$ is the homomorphism such that $v(a_r) = u^{c_r}$, $r \in R$. d_χ is called the *generic degree* associated with χ.

PROPOSITION 2. *For each χ, and f^*, as above, d_χ belongs to the specialization ring of f^*, and $f^*(d_\chi) = \deg(\zeta_{\chi,f^*})$.*

PROPOSITION 3. *Suppose χ is rational, in the sense that $\chi(a_w) \in \mathfrak{o}$ for all $w \in W$. Then $d_\chi \in Q[u]$, a polynomial with rational coefficients.*

To obtain a criterion for rationality, we proceed as follows. Let $J \subset R$, and let χ be an irreducible character of $A^{\bar{K}}$. Let $e_J = \sum_{w \in W_J} a_w$, and $E_J = v(e_J)$. Then $\chi(e_J) = m_J(\chi) E_J$ for some nonnegative integer $m_J(\chi)$. For each specialization $f : u \to q$ or 1, $m_J(\chi) = (1^G_{G_J}, \zeta_{\chi, f^*})$ where G_J is the parabolic subgroup $BW_J B$ (see [4, §7]). Suppose χ is not rational; then $\chi \ne \chi^\sigma$ for some field automorphism σ of $\bar{K}$ over K. Moreover, $m_J(\chi) = m_J(\chi^\sigma)$ for all subsets $J \subset R$. We can now state the following result.

THEOREM 1. *Suppose χ is an irreducible character of $A^{\bar{K}}$ which is such that for all other irreducible characters $\chi' \ne \chi$, $m_J(\chi) \ne m_J(\chi')$ for some $J \subset R$. Then χ is a rational character of $A^{\bar{K}}$, i.e. $\chi(a_w) \in \mathfrak{o}$, $w \in W$.*

It is sufficient to check the rationality criterion for the Weyl groups. Using the character tables of Kondo and Frame for the exceptional groups, the following result can be proved.

THEOREM 2. *Suppose (W, R) is of type A_n $(n \ge 1)$, B_n (or C_n) $(n \ge 2)$, D_n $(n \ge 4)$, G_2, F_4, E_6, E_7, or E_8. Then every irreducible character of the generic algebra $A(u)^{\bar{K}}$ is rational, with the possible exception of the two irreducible characters of degree $512 = 2^9$, in case (W, R) is of type E_7.*

We turn now to the question of the rationality of the characters $\zeta \in 1^G_B$ on the whole group $G = G(q)$. The rationality of the irreducible complex characters of W is a well-known result, for W of type A_n, B_n, C_n, D_n, E_n, F_4, and G_2.

Let $G = G(q)$ be as above. An irreducible character $\zeta \in 1^G_B$ is said to be of *parabolic type* in case $(\zeta, 1^G_P) = 1$ for some parabolic subgroup P of G (see [4]). Similarly, characters of W of parabolic type are defined.

EXAMPLE. For W of type A_n $(n \ge 1)$, all irreducible characters of W are of parabolic type, and using the previous results about multiplicities, it follows that all irreducible characters in 1^G_B, for $G = G(q)$ of type A_n, are of parabolic type.

Using a result of Janusz [5], we have the following result.

THEOREM 3. *Let χ be an irreducible character of the generic algebra $A^{\bar{K}}$, as above, and let $f : u \to q$ denote a specialization for some characteristic power q, and $f_0 : u \to 1$. Then the character ζ_{χ, f^*} of $G(q)$ is of parabolic type if and only if ζ_{χ, f_0} is of parabolic type. In that case, let*

$$e(\zeta_{\chi, f^*}) = \frac{\deg \zeta_{\chi, f^*}}{|G|} \sum_{x \in G} \zeta_{\chi, f^*}(x^{-1}) x,$$

and

$$e_P = \frac{1}{|P|} \sum_{x \in P} x,$$

for the parabolic subgroup P such that $(\zeta_{\chi,f^*}, 1_P^G) = 1$. Then $e(\zeta_{\chi,f^*})e_P$ (see [3] for an explicit formula for this element) is a primitive idempotent in the group algebra of G affording the character ζ_{χ,f^*}. In particular, if χ is rational on $A^{\bar{K}}$, then ζ_{χ,f^*} is rational on G.

To handle characters $\zeta \in 1_B^G$ which may not be of parabolic type, we modify the above procedure as follows. Let $\varepsilon: A \to \mathfrak{o}$ be the homomorphism of the generic ring such that $\varepsilon(a_w) = (-1)^{l(w)}$, $w \in W$. Now, let $P = BW_J B$ be a parabolic subgroup, and let

$$\varepsilon_J = \frac{u^{N_J}}{E_J} \sum_{w \in W_J} \frac{(-1)^{l(w)}}{u^{l(w)}} a_w,$$

where N_J is the number of positive roots of W_J, and $E_J = \sum_{w \in W_J} v(a_w)$.

PROPOSITION 4. *Let f, f_0 be as in Theorem 4. Then*

$$\varepsilon_{J,f_0} = \frac{1}{|W_J|} \sum_{w \in W_J} (-1)^{l(w)} w, \quad \text{and} \quad \varepsilon_{J,f} = \frac{q^{N_J}}{f(E_J)} \sum_{w \in W_J} \frac{(-1)^{l(w)}}{q^{l(w)}} a_{wf}$$

are primitive idempotents in the group algebras of W_J and P, affording the alternating character ε_J and the Steinberg character σ_J [2] of W_J and P respectively. Moreover, if χ is an irreducible character of $A^{\bar{K}}$, then the multiplicities $(\zeta_{\chi,f_0}, \varepsilon_J^W)$ and $(\zeta_{\chi,f^*}, \sigma_J^G)$ coincide.

THEOREM 4. *Let the notation be as in Proposition 4. If $(\zeta_{\chi,f_0}, \varepsilon_J^W) = 1$ for some J, then $e(\zeta_{\chi,f^*}) \varepsilon_{J,f}$ is a primitive idempotent in the group algebra of G affording the character ζ_{χ,f^*}, and can be expressed as a linear combination of the elements a_{wf} with coefficients in $Q(\chi_f(a_{wf}); w \in W)$. In particular, if χ is rational on $A^{\bar{K}}$, then ζ_{χ,f^*} is rational on G.*

Combining Theorems 2, 3 and 4 and making further use of the tables of Kondo and Frame, we obtain:

THEOREM 5. *Let W be of type G_2, F_4, E_6, E_7, E_8, and let G be a group belonging to a system of type W. Then every irreducible complex character ζ of G such that $(\zeta, 1_B^G) > 0$ has the property that $\zeta(g) \in \mathbf{Z}$ for all $g \in G$, with the possible exception of two characters in case W is of type E_7 (see Theorem 2) and is afforded by a rational representation of G.*

As a corollary to the proof of this result, we obtain the theorem that for W as in Theorem 6, all irreducible representations of the Weyl Group W, without exception, are realizable in the rational field, a result obtained earlier and by different methods by Benard [1].

References

1. M. Benard, *On the Schur indices of the characters of the exceptional Weyl groups*, Ph.D. Dissertation, Yale University, New Haven, Conn., 1969.

2. C. W. Curtis, *The Steinberg character of a finite group with a (B,N)-pair*, J. Algebra **4** (1966), 433–441. MR **34** #1406.

3. C. W. Curtis and T. V. Fossum, *On centralizer rings and characters of representations of finite groups*, Math. Z. **107** (1968), 402–406. MR **38** #5946.

4. C. W. Curtis, N. Iwahori and R. Kilmoyer, *Hecke algebras and characters of parabolic type of finite groups with (B,N)-pairs* (to appear).

5. G. J. Janusz, *Primitive idempotents in group algebras*, Proc. Amer. Math. Soc. **17** (1966), 520–523. MR **33** #2733.

UNIVERSITY OF OREGON

Types of Blocks of Representations of Finite Groups

Richard Brauer

Let G be a group of finite order g. Let p be a fixed prime. We are concerned with the behavior of the characters χ of the complex irreducible representations of G with regard to the prime p. The basic formulas are

(1) $$\chi(\pi\rho) = \sum_{\phi^\pi} d^\pi(\chi,\phi^\pi)\phi^\pi(\rho)$$

where π is a p-element of G, ρ is a p-regular element of the centralizer $C(\pi)$ of π, ϕ^π ranges over the modular irreducible characters of $C(\pi)$ for p, and the decomposition numbers $d^\pi(\chi,\phi^\pi)$ are algebraic integers of a certain cyclotomic field which do not depend on ρ. Since characters are class functions, it suffices to take π in a system of representatives for the conjugacy classes of p-elements of G and, for each π, to take ρ in a system of representatives for the p-regular conjugacy classes of $C(\pi)$. Then the full character table of G can be obtained from (1).

We may compare (1) with the corresponding formulas for $C(\pi)$. We take here $\pi = 1$ in the argument of the character. We have

(2) $$\tilde{\chi}(\rho) = \sum_{\phi^\pi} d_0(\tilde{\chi},\phi^\pi)\phi^\pi(\rho)$$

where the $\tilde{\chi}$ are the irreducible characters of $C(\pi)$ and where the $d_0(\tilde{\chi},\phi^\pi)$ are the decomposition numbers of $C(\pi)$. If ϕ_1^π,ϕ_2^π are two modular irreducible characters of $C(\pi)$, then

(3) $$\sum_\chi \overline{d^\pi(\chi,\phi_1^\pi)}\, d^\pi(\chi,\phi_2^\pi) = \sum_{\tilde{\chi}} \overline{d_0^\pi(\tilde{\chi},\phi_1^\pi)}\, d_0^\pi(\tilde{\chi},\phi_2^\pi) = c^\pi(\phi_1^\pi,\phi_2^\pi).$$

The $c^\pi(\phi_1^\pi,\phi_2^\pi)$ are rational integers, the Cartan invariants of $C(\pi)$. On the other

---

AMS 1970 *subject classifications*. Primary 20C15.

hand, if π' is a p-element of G not conjugate to π and if $\phi_2^{\pi'}$ is a modular irreducible character of $C(\pi')$, we have

(4) $$\sum_\chi \overline{d^\pi(\chi,\phi^\pi)}\, d^{\pi'}(\chi,\phi^{\pi'}) = 0.$$

Let B now denote a fixed p-block of G. As is well known, it suffices for $\chi \in B$ to let ϕ^π in (1) range over the modular irreducible characters in blocks b of $C(\pi)$ which satisfy the relation $b^G = B$. The set of these blocks b of $C(\pi)$ will be denoted by $\mathrm{BL}(C(\pi),B)$.

If $b \in \mathrm{BL}(C(\pi),B)$ and if $\tilde\chi \in b$, only modular irreducible characters ϕ^π of b occur in (2) with nonzero coefficients. We have $d^\pi(\chi,\phi^\pi) = 0$, if the irreducible character χ of G belongs to a block different from $B = b^G$ while ϕ^π is a modular character in b. It follows that it suffices in (3) to let χ range over B and $\tilde\chi$ over b. An analogous remark applies in (4).

Let B be a given block. We shall consider only characters $\chi \in B$ and ordinary characters $\tilde\chi$ and modular characters ϕ^π of blocks $b \in \mathrm{BL}(C(\pi),B)$ for p-elements π of G. With B, there is associated a class of conjugate p-groups of G, the defect groups D of B. If $|D| = p^d$, d is the defect of B. For $b \in \mathrm{BL}(C(\pi),B)$, a defect group D_0 of b can be chosen as a subgroup of a defect group D of B. In particular, the defect d_0 of b is at most equal to the defect d of B.

It is not known whether or not for given defect d the coefficients $d^\pi(\chi,\phi^\pi)$ in (1) always belong to a finite set of complex numbers $M(p^d)$ which does not depend on the group G or on the particular choice of the p-block B of G of defect d. If this should be true (as it is for instance when $d = 1$), then we would have only finitely many possibilities for the matrices

(5) $$T^\pi(B,b) = (d^\pi(\chi,\phi^\pi))$$

where the "row index" χ ranges over the irreducible characters of B and the column index ϕ^π over the modular irreducible characters of b, ($b \in \mathrm{BL}(C(\pi),B)$). If we know which of the finitely many possibilities applies for each b, and if in addition we have a method of constructing the irreducible modular characters of the group $C(\pi)$, we could determine the $\chi(\pi\rho)$ in (1). Unfortunately, no explicit construction of the modular irreducible characters of an arbitrary given group is known.

In order to avoid these difficulties, we have already given in [1, §5] a modification of our method. We shall modify this method still further, and this will allow us to derive much more precise results. As in [1], we consider a set of expressions which are obtained from the set of modular irreducible characters ϕ^π of b by a linear transformation with integral rational coefficients and the determinant ± 1. Any such set of expressions will be called a *basic set* $W(b)$ of b. It is then immediate that we have again formulas (1) if for each $b \in \mathrm{BL}(C(\pi),B)$ we have a basic set $W(b)$ for b and if ϕ^π ranges over the union of these $W(b)$. Again, we have formulas (2) with ϕ^π now also ranging over $W(b)$. The Cartan invariants $c^\pi(\phi_1^\pi,\phi_2^\pi)$ can be defined for $\phi_1^\pi,\phi_2^\pi \in W(b)$ and (3) holds again. As in [1], the starting point is the following result.

LEMMA 1. *Let p be a prime and $d \geq 0$ a rational integer. There exist functions $f(p,d)$ and $g(p,d)$ of p and d with the following properties*: 1. *If B is a p-block of defect d of an arbitrary finite group G, the number $k(B)$ of ordinary irreducible characters χ in G is at most equal to $f(p,d)$.* 2. *There exists a basic set $W(B)$ for B such that the corresponding Cartan invariants $c(\phi_1,\phi_2)$ all satisfy the inequality $|c(\phi_1,\phi_2)| \leq g(p,d)$.*

This can be deduced fairly easily from known results. We sketch here still another proof which is often preferable in concrete situations. We use induction on d. The case $d = 0$ is trivial. Suppose then that $d > 0$ and that the result is known for blocks of defect smaller than d. If π is a p-element of G which is not conjugate to an element of the defect group D of B, the set $\mathrm{BL}(C(\pi),B)$ is empty, and π need not be considered. Choose a set Π of representatives for the conjugacy classes of G which meet D taking $\Pi \subseteq D$. If $\pi \in \Pi$ and $\pi \neq 1$, a block b of $C(\pi)$ corresponds in a natural fashion to a block $\bar{b}$ of $C(\pi)/\langle\pi\rangle$. The modular irreducible characters of b and of $\bar{b}$ can be identified. If the Cartan invariant of $\phi_1,\phi_2 \in \bar{b}$ has the value c for $\bar{b}$, the corresponding Cartan invariant of ϕ_1,ϕ_2 for b has the value $p^r c$, where p^r is the order of π. Likewise, if $\bar{b}$ has defect d_1 then b has defect $d_1 + r$. In particular, if $b \in \mathrm{BL}(C(\pi),B)$, then $d_1 < d_1 + r \leq d$ and our induction hypothesis applies to $\bar{b}$. It follows that there exists a basic set W of $\bar{b}$ for which the absolute values of the Cartan invariants lie below a bound depending only on p^d. The same still remains true if we consider W as a basic set $W(b)$ of b. It then follows from (3) that the decomposition numbers $d^\pi(\chi,\phi)$ belong to a finite set of complex numbers depending only on p and d. Moreover, if k_0 is the number of irreducible characters $\chi \in B$ such that $d^\pi(\chi,\phi) \neq 0$ for some $\phi \in W$, then k_0 lies below a bound depending only on p and d. If we choose $\pi \neq 1$ such that π is conjugate to an element of $Z(D)$, then there exist $b \in \mathrm{BL}(C(\pi),B)$ of defect d. For such b, actually $k_0 = k(B)$, cf. [2]. This yields the statement 1 of Lemma 1. If we define $T^\pi(B,b)$ as in (5), but use the basic set $W(b)$ instead of the modular irreducible characters of b, we now see that there are only finitely many possibilities for each of the matrices $T^\pi(B,b)$ with $\pi \neq 1$. For $\pi = 1$, we have $C(\pi) = G$, $\mathrm{BL}(C(\pi),B) = \{B\}$. If we set $T = T^1(B,B)$ for an arbitrary choice of a basic set $W(B)$, then T has the following properties: The number of columns is equal to the number $l(B)$ of modular irreducible characters of B. The coefficients of T are rational integers. By (4), the columns of T are orthogonal to the columns of all $T^\pi(B,b)$ with $\pi \neq 1$. Finally, there exist left inverses Y of T (i.e., matrices Y for which YT is the unit matrix of degree $l(B)$) and Y can be chosen with integral rational coefficients. Conversely, if T has all these properties, T is the matrix $T^1(B,B)$ for a suitable choice of the basic set $W(B)$ of B. It is now clear that $W(B)$ can be chosen such that the decomposition numbers $d^1(\chi,\phi)$ for $\chi \in B$, $\phi \in W(B)$ belong to a finite set of complex numbers and this sets the second part of Lemma 1 into evidence.

Our method also yields

LEMMA 2. *There exists a finite set $M(p,d)$ of complex numbers depending only on*

p and d such that if B is a p-block of defect d, if π is a p-element of G, and if for each $b \in \mathrm{BL}(C(\pi),B)$ a basic set $W(b)$ is used for which the absolute values of the Cartan invariants are at most equal to $g(p,d_0)$, (d_0 the defect of b), then all decomposition numbers $d^\pi(\chi,\phi^\pi)$ belong to $M(p,d)$.

We now come to the definition of the *type* of a block. If b and π are as before and if $W(b)$ is a basic set for b, then as in (5), we form the matrices $T^\pi(B,b) = (d^\pi(\chi,\phi^\pi))$ with $\chi \in B$ as row index and $\phi^\pi \in W(b)$ as column index. In addition, we form the matrix

$$(6) \qquad T_0(b) = (d_0(\tilde{\chi},\phi^\pi))$$

of the ordinary decomposition numbers $d_0(\tilde{\chi},\phi^\pi)$ of b with $\tilde{\chi} \in b$ serving as row index and $\phi^\pi \in W(b)$ as column index.

DEFINITIONS. 1. Let B be a p-block of defect d of a finite group G. Let π be a p-element of G. We say that B is of a *given type with regard to a p-element π of G*, if for each $b \in \mathrm{BL}(C(\pi),B)$, the pair of matrices

$$(7) \qquad T^\pi(B,b), \qquad T_0(b)$$

cf. (5), (6), formed with regard to some basic set $W(b)$, is given. No further information concerning $W(b)$ is required.

2. We say that B is of *known type*, if it is of known type for all p-elements π of G.

The following result is an immediate consequence of the lemmas.

THEOREM 1. *For given p and d, there exist only finitely many types of p-blocks of defect d (for the category of all finite groups).*

We shall now see what we can say about the values of the irreducible characters χ of a p-block B of defect d of G, if for a set Π_0 of p-elements π of G, we know the type of B with regard to π for $\pi \in \Pi_0$ and if we have additional information concerning the centralizers $C(\pi)$ for these π. We introduce some notation. If H is a finite group, we denote by $\mathrm{CL}(H)$ the set of conjugacy classes of H and by $\mathrm{RCL}(H)$ the subset of p-regular conjugacy classes of H. If H is a subgroup of the finite group G, each $L \in \mathrm{CL}(H)$ belongs to a unique $K \in \mathrm{CL}(G)$ which we denote by $K = L^G$. We thus have a mapping of $\mathrm{CL}(H)$ into $\mathrm{CL}(G)$ which maps $\mathrm{RCL}(H)$ into $\mathrm{RCL}(G)$.

THEOREM 2. *Let G be a finite group and let D of order p^d be a p-subgroup. Let Π denote a set of representatives for the conjugacy classes of G which meet D, chosen with $\Pi \subseteq D$, and let Π_0 be a nonempty subset. Let B be a block of G with the defect group D. Suppose we have the following information:*

(i) *For each $\pi \in \Pi_0$ we know the type of B with regard to π.*

(ii) *For each $\pi \in \Pi_0$ and $L \in \mathrm{RCL}(C(\pi))$, we know $|L|$; we know $L^G \in \mathrm{RCL}(G)$; we know for which of these L the class L^G has defect less than d, and for which L the class L^G has the defect group D.*

(iii) *For* $\pi \in \Pi_0$ *we know the values of the irreducible characters of* $C(\pi)$ *in blocks* b *of defect at most* d *for the classes* $L \in \mathrm{RCL}(C(\pi))$.

(iv) *For at least one* $\pi_0 \in \Pi_0$ *we know a block* b_0 *of* $C(\pi_0)$ *for which* $b_0^G = B$.

Then we can determine the values of the irreducible characters χ *in* B *for all elements of the form* $\sigma = \pi\rho$ *with* $\pi \in \Pi_0$ *and p-regular* $\rho \in \Pi_0$ *(assuming that we know to which* $L \in \mathrm{RCL}(C(\pi))$ *the element* ρ *belongs).*

The proof is based on the formula (1) with $\pi \in \Pi_0$ and with the ϕ^π taken as the members of the basic sets $W(b)$ which occur in the definition of the type of B with regard to π. It follows from the main theorems on blocks that, with the given information, we can identify the blocks $b \in \mathrm{BL}(C(\pi),B)$. Then the $d^\pi(\chi,\phi^\pi)$ are the coefficients of the matrix $T^\pi(B,b)$ in (5), and since we know the type of B with regard to π, the $d^\pi(\chi,\phi^\pi)$ are known. Again, if we write (2) for b with ϕ^π taken in $W(b)$, the coefficients $d_0(\tilde{\chi},\phi^\pi)$ are the coefficients of $T_0(b)$ in (6) and hence are known. The matrix $T_0(b)$ has a left inverse and, of course, such a left inverse can actually be found. It then follows from (2) that each $\phi^\pi \in W(b)$ can be written explicitly as a linear combination of the $\tilde{\chi} \in b$. Since the values of the $\tilde{\chi}$ for classes $L \in \mathrm{RCL}(C(\pi))$ are known, we can actually find the $\phi^\pi \in W(b)$. Now (1) yields the desired result.

We mention some consequences. The assumptions and the notation will be as in Theorem 2.

REMARK 1. It follows from (4) that if σ is an element of G whose p-factor is not conjugate to an element of Π_0 then, for all $\pi \in \Pi_0$ and $\phi^\pi \in W(b)$ with $b \in \mathrm{BL}(C(\pi),B)$,

$$(8) \qquad \sum_{\chi \in B} d^\pi(\chi,\phi^\pi)\chi(\sigma) = 0.$$

Since the $d^\pi(\chi,\phi^\pi)$ are known we have then linear equations for the values $\chi(\sigma)$. Actually, any such linear equation which holds for all σ considered here is a linear combination of relations (8).

If $1 \notin \Pi_0$, we may take $\sigma = 1$ in (8) and obtain linear relations between the degrees $\chi(1)$ of the characters $\chi \in B$.

REMARK 2. Let σ be a p-regular element of G. Suppose that H is a p-subgroup of $C(\sigma)$ such that every $\pi \neq 1$ in H is conjugate to an element of Π_0. Since then $\chi(\pi\sigma)$ can be found from Theorem 2, we can determine the value of $\chi(\sigma)$ (mod $|H|$).

REMARK 3. Assume that $\Pi_0 = \Pi - \{1\}$. Let P denote a Sylow p-subgroup of G. Suppose that for each $\pi \in \Pi_0$, we know the number of elements of P which are conjugate to π. A method similar to that in Remark 2 allows us to determine the value of $\chi(1)$ (mod $|P|$) for $\chi \in B$.

REFERENCES

1. R. Brauer, *Some applications of the theory of blocks of characters of finite groups.* I, J. Algebra **1** (1964), 152–167; IV (to appear). MR **29** #5920.

2. ———, *On blocks and sections in finite groups.* II, Amer. J. Math. **90** (1968), 895–925. MR **39** #5713.

HARVARD UNIVERSITY.

Modular Representations of Some Finite Groups

N. Burgoyne

The groups that are of interest here are the finite Chevalley groups. These are the analogues, over a finite field, of the complex simple Lie groups. By and large they account for most of the known simple groups.

The question studied in the following sections is how to describe explicitly the irreducible representations of a Chevalley group over a field having the *same* characteristic as the defining field of the group. Due principally to the work of C. W. Curtis and R. Steinberg, there now exists a rather complete and elegant theory which gives a general description of these representations. An excellent exposition of these results is contained in [1]. In many ways it is not too different from the classical work of E. Cartan.

As in the characteristic zero case one can characterize any finite-dimensional irreducible representation by its unique highest weight. The properties of the representation, for example its Brauer character, are readily described once the multiplicity of each sub-weight is known. At characteristic zero this multiplicity may be computed by several methods, the formulas of Freudenthal and Kostant being the most practical (see, for example, [2, Chapter VIII]). For nonzero characteristic no such formulas are known although useful results are contained in papers by T. A. Springer [3], J. Humphries [4], and W. Wong [5]. Otherwise, except for some examples of groups with rank ≤ 2, very little is known about the multiplicities and hence about the representations.

In §1, after outlining our notation, we give an *explicit* method of computing the above-mentioned multiplicities. The only fault with this method is that it is tedious. In §2 is a result which allows one to compute with ease the multiplicities of certain rather special sub-weights.

AMS 1970 *subject classifications.* Primary 20C20, 20D05; Secondary 17B20.

To conclude this introduction the author would like to thank J. Humphries for several very helpful discussions. He also wishes to thank the National Science Foundation for support, during a portion of this work, via NSF Grant GJ-821.

1. Let G be a complex simple Lie algebra of rank l and U its universal enveloping algebra. We use the following notation:

Σ is a system of roots for G;
Π is a system of simple roots in Σ, call them $\alpha_1, \alpha_2, \ldots, \alpha_l$;
Σ^+ is the positive roots in Σ relative to Π;
Λ denotes the set of all dominant integral weights relative to Π. Let $\lambda_1, \lambda_2, \ldots, \lambda_l$ denote the fundamental weights in Λ.

The elements of all the above sets may be envisioned as vectors in the same l-dimensional Euclidean vector space V over the rational numbers. Let the inner product in V be denoted by (α,β) and for $\beta \neq 0$ put $\langle \alpha,\beta \rangle = 2(\alpha,\beta)/(\beta,\beta)$. For $\alpha, \beta \in \Sigma$, $\langle \alpha,\beta \rangle$ is an integer.

Recall that $\langle \lambda_i, \alpha_j \rangle = \delta_{ij}$ and any $\lambda \in \Lambda$ has the form $\lambda = \sum m_i \lambda_i$ where the m_i are nonnegative integers. For use in §2 we define the weight $\delta = \sum \lambda_i$.

The elements of Λ may be partitioned into disjoint "chains." If $\lambda, \mu \in \Lambda$ then they are in the same chain if and only if $(\lambda - \mu)$ is an integral combination of elements of Σ. The number of distinct chains is equal to the determinant of the Cartan matrix (c_{ij}) where $c_{ij} = \langle \alpha_j, \alpha_i \rangle$.

If λ, μ are in the same chain, then we say that μ is a sub-weight of λ, and write $\mu < \lambda$, if $(\lambda - \mu)$ is nonzero and is an integral combination of elements of Σ^+ with nonnegative coefficients. We denote by $P(\lambda - \mu)$ the number of distinct ways of writing $(\lambda - \mu)$ in this form. The relation $\mu < \lambda$ induces a partial ordering on the elements of a chain. Each chain has a unique smallest weight.

Suppose that in G we have a Chevalley basis $\{e_\alpha, f_\alpha, h_{\alpha_i}\}$ with $\alpha \in \Sigma^+$ and $i = 1, \ldots, l\}$. If $\beta \in \Sigma$ and $\beta = \sum n_i \alpha_i$, then define $h_\beta = \sum n'_i h_{\alpha_i}$ where $n'_i = n_i(\alpha_i, \alpha_i)/(\beta,\beta)$. In this notation we have the following commutators:

$$[h_\beta, e_\alpha] = \langle \alpha,\beta \rangle e_\alpha, \qquad [h_\beta, f_\alpha] = -\langle \alpha,\beta \rangle f_\alpha, \qquad [e_\alpha, f_\alpha] = h_\alpha.$$

Put $N = |\Sigma^+|$ and let $S = \{s_1, s_2, \ldots, s_N\}$ denote any sequence of nonnegative integers. For some fixed ordering in $\Sigma^+ = \{\beta_1, \beta_2, \ldots, \beta_N\}$, define in U the element

$$e_S = \frac{e_{\beta_1}^{s_1}}{s_1!} \cdot \frac{e_{\beta_2}^{s_2}}{s_2!} \cdots \frac{e_{\beta_N}^{s_N}}{s_N!}.$$

The element f_S is defined in a similar manner. In the vector space V let $\sigma_S = \sum s_i \beta_i$. The number of distinct $S, S', \ldots$ satisfying $\sigma_S = \sigma_{S'} = \cdots = \sigma$ is clearly $P(\sigma)$.

Now, given $\lambda \in \Lambda$, one constructs a U-module F_λ, generated by a vector v_+ and with a basis $f_S v_+$ where S runs over all sequences of nonnegative integers, by defining $e_\alpha v_+ = 0$ and $h_\alpha v_+ = \langle \lambda, \alpha \rangle v_+$. Let $v \in F_\lambda$ be any linear combination of $f_S v_+, f_{S'} v_+, \ldots$ all satisfying $\sigma_S = \sigma_{S'} = \cdots = \sigma$. If $\sigma_T = \sigma$, then $e_T v = a_T v_+$

where a_T is a number. Define N_λ to be the subspace of F_λ generated by those v for which $a_T = 0$ for all such T. N_λ is invariant under U and $M_\lambda = F_\lambda/N_\lambda$ is irreducible and finite dimensional.

If $\mu \in \Lambda$ and $\mu < \lambda$ then the multiplicity of μ in λ is the dimension of the image in M_λ of the subspace in F_λ spanned by all $f_S v_+$ with $\sigma_S = \lambda - \mu$. Hence this multiplicity is just the rank of the matrix (a_{TS}), of degree $P(\lambda - \mu)$, where

$$e_T f_S v_+ = a_{TS} v_+$$

and $\sigma_T = \sigma_S = \lambda - \mu$.

Now by a theorem of B. Kostant [6] the numbers a_{TS} are integers. As described in [1], this allows one to "reduce" M_λ modulo p giving a representation $\overline{M}_\lambda$ of the Lie algebra $\overline{G}$, corresponding to G but defined over the field with p elements. In general $\overline{M}_\lambda$ is not irreducible. However, if λ is a restricted weight, that is, if $\lambda = \sum m_i \lambda_i$ and each $m_i < p$, then $\overline{M}_\lambda$ contains a unique highest composition factor $\overline{K}_\lambda$. The work of Curtis and Steinberg implies that to determine the sub-weight structure of any modular irreducible representation of a Chevalley group corresponding to $\overline{G}$, we need only know it for the modules $\overline{K}_\lambda$.

In F_λ define the subspace $N_\lambda(p)$ by replacing, in the definition of N_λ, the condition $a_T = 0$ by $a_T \equiv 0 \pmod{p}$. Let $\overline{N_\lambda(p)}$ denote its image in $\overline{M}_\lambda$. Then $\overline{K}_\lambda = \overline{M}_\lambda/\overline{N_\lambda(p)}$. Thus we have the following result.

PROPOSITION. *Let $\mu, \lambda \in \Lambda$ with λ restricted and $\mu < \lambda$. The multiplicity of μ in $\overline{K}_\lambda$ is equal to the number of elementary divisors of the integral matrix (a_{TS}) which are prime to p.*

Thus to compute the multiplicities one must compute the numbers a_{TS}. This requires the full commutation relations in G but is otherwise simple arithmetic.

As an example consider the 6×6 symplectic group C_3 and the high weight $\lambda = \lambda_1 + \lambda_2$ (where α_1 and α_2 are short). M_λ has degree 64 and has two dominant sub-weights,

$$\mu_1 = \lambda_3, \quad \lambda - \mu_1 = \alpha_1 + \alpha_2, \quad P(\lambda - \mu_1) = 2,$$
$$\mu_2 = \lambda_1, \quad \lambda - \mu_2 = \alpha_1 + 2\alpha_2 + \alpha_3, \quad P(\lambda - \mu_2) = 7.$$

For μ_1 the 2×2 matrix has elementary divisors 1,3. For μ_2 the 7×7 matrix has as nonzero elementary divisors 1,1,1,21. Thus $\overline{M}_\lambda$ is irreducible unless $p = 3$ or 7. One also sees that $\overline{K}_\lambda$ has degree 50 when $p = 3$ and degree 58 when $p = 7$.

Provided that the value of $P(\lambda - \mu)$ is small, say less than 10, this method is quite feasible for hand computation. Use of a machine can extend this bound considerably, but one is still limited. For example, in one of the fundamental weights of E_8 the multiplicity (at characteristic 0) of the sub-weight $\mu = 0$ exceeds 10^6 (see [7]). This prompts one to search for other approaches.

2. Let λ be a restricted weight in Λ. Then $\overline{K}_\lambda$ occurs as the unique top composition factor in $\overline{M}_\lambda$. Since the multiplicity of any sub-weight of $\overline{M}_\lambda$ is known (by

the formula for characteristic zero), another means of computing the multiplicities in $\bar{K}_\lambda$ is to determine which sub-weights of λ serve as high weights for composition factors in $\bar{M}_\lambda$. By induction one may suppose that the multiplicities in these factors are known. Note that a factor need not have a *restricted* high weight, but in this case the "tensor product" theorem (see [1]) allows one to calculate the multiplicities.

The following result gives an easily used criterion for determining, in certain cases, when μ is the high weight of a factor of $\bar{M}_\lambda$. Roughly speaking the criterion applies if $(\lambda - \mu)$ is small. Let

$$\lambda = \sum m_i \lambda_i \quad \text{and} \quad \lambda - \mu = \sum n_i \alpha_i;$$

then the m_i and n_i are nonnegative integers.

PROPOSITION. *Suppose that*
(a) $n_i \leq m_i$;
(b) $\lambda - \mu = m\alpha$ *for some* $\alpha \in \Sigma^+$ *and integer* m, *then* $\bar{M}_\lambda$ *has a composition factor with high weight* μ *if and only if*

$$\langle \lambda + \delta, \alpha \rangle \equiv m \pmod{p}.$$

Furthermore, if it exists, this is the only factor with μ as its highest weight.

The necessity of this condition is quickly proved and in fact is closely related to the more general results of T. Springer [3, Proposition 4.2] and J. Humphries [4]. The sufficiency requires a detailed discussion of certain equations in U and in its quotient ring. The assumption (a) is crucial for this part of the proof. The assumption (b) is rather minor and it seems likely that the result can be extended in such a way that (b) can be dropped. The full proof will appear elsewhere.

This proposition is not usually sufficient to give a complete determination of the multiplicities of all the sub-weights of λ. However, there are some cases where it does. As an example consider the 4×4 linear group A_3 and its representation with high weight $\lambda = \lambda_1 + \lambda_2 + \lambda_3$. M_λ has degree 64 and has three dominant sub-weights

$$\mu_1 = 2\lambda_1, \qquad \lambda - \mu_1 = \alpha_2 + \alpha_3, \qquad P(\lambda - \mu_1) = 2,$$
$$\mu_2 = 2\lambda_3, \qquad \lambda - \mu_2 = \alpha_1 + \alpha_3, \qquad P(\lambda - \mu_2) = 2,$$
$$\mu_3 = \lambda_2, \qquad \lambda - \mu_3 = \alpha_1 + \alpha_2 + \alpha_3, \qquad P(\lambda - \mu_3) = 4.$$

Conditions (a) and (b) are clearly satisfied and $m = 1$ in each case. The proposition then shows that μ_1 and μ_2 are high weights of factors when $p = 3$ and μ_3 when $p = 5$. From this it is easily seen that $\bar{K}_\lambda$ has degree 44 for $p = 3$ and degree 58 for $p = 5$. $\bar{M}_\lambda$ is irreducible for all other p.

References

1. R. Steinberg, *Lectures on Chevalley groups*, Yale University, Dept. of Mathematics, New Haven, Conn., 1967.

2. N. Jacobson, *Lie algebras*, Interscience Tracts in Pure and Appl. Math., no. 10, Interscience, New York, 1962. MR **26** #1345.

3. T. A. Springer, *Weyl's character formula for algebraic groups*, Invent. Math. **5** (1968), 85–105. MR **37** #2763.

4. J. Humphries, *Modular representations of classical Lie algebras and semi-simple groups* (to appear).

5. W. Wong, *Representations of Chevalley groups in characteristic p* (to appear).

6. B. Kostant, *Groups over Z*, Proc. Sympos. Pure Math., vol. 9, Amer. Math. Soc., Providence, R.I., 1966, pp. 90–98. MR **34** #7528.

7. H. Freudenthal, *Zur Berechnung der Charaktere der halbeinfachen Lieschen Gruppen*. II, Nederl. Akad. Wetensch. Proc. Ser. A **57** = Indag. Math. **16** (1954), 487–491. MR **16**, 673.

UNIVERSITY OF CALIFORNIA, SANTA CRUZ

Some Connections Between Clifford Theory and the Theory of Vertices and Sources

Edward Cline[1]

This note provides statements for the main results of [1], [2], [3] which establish some connections between the recent results of Dade [5], [6], [7] on Clifford theory and the theory of vertices and sources as developed by Green [9], [10], [11].

I. **Minimal vertices.** Let $\mathfrak{D}$ be a complete local domain with maximal ideal $\mathfrak{P}$ whose quotient field $\mathfrak{K}$ has characteristic zero, and whose residue class field $\mathfrak{D}/\mathfrak{P}$ is algebraically closed of characteristic $p > 0$. Let G be a finite group and assume all $\mathfrak{D}(G)$-modules are right modules which are free and finitely generated as $\mathfrak{D}$-modules.

If χ is an absolutely irreducible $\mathfrak{K}$-character of G, the following question appears in Feit's notes [8] on modular theory:

(1) Does there exist an $\mathfrak{D}(G)$-module M affording χ with vertex B such that if $B \leqq P = S_p(G)$, then $[P:B] = \chi(1)_p$ where $\chi(1)_p$ denotes the p-part of the degree of χ?

The answer to this question is no because of the following example. Let $p = 2$, $G = \mathrm{SL}(2,3)$, and let χ be a faithful character of G of degree 2. If M affords χ, it is easy to see that the vertex of M must be a S_2-subgroup of G. In view of this and other examples, we modify (1) as follows:

Let $\mathscr{V}_\chi$ denote the collection of all vertices of $\mathfrak{D}(G)$-modules M affording χ. Let $|G|_p$ denote the p-part of $|G|$, and set

$$\alpha_\chi = \max \{|G|_p/|B| \,|\, B \in \mathscr{V}_\chi\}.$$

Then ask

(1')   How can we compute, or characterize, α_χ?

AMS 1970 *subject classifications*. Primary 20C20, 20C05.

[1] This research was supported in part by NSF grant GP 9649.

So far, we can answer this question only in the following special situation:

Let H be a normal p-subgroup of G, and assume $\chi_H = \phi$ is absolutely irreducible. Then define

$$(2) \quad \mathcal{U}(\phi) = \left\{ B \leq H \;\middle|\; \begin{array}{l} \text{(a) } N_G(B) \text{ covers } G/H, \\ \text{(b) There exists a character } \mu \text{ of } B \text{ such that} \\ \quad \text{(i) } \mu^H = \phi, \text{ and} \\ \quad \text{(ii) the stabilizer } \Sigma_B \text{ of } \mu \text{ in } N_G(B) \text{ covers } G/H. \end{array} \right\}$$

THEOREM 1. *Under the hypothesis mentioned above,*

$$\alpha_\chi = \max \{[H:B] \mid B \in \mathcal{U}(\phi)\}.$$

In fact, we show that the minimal elements of $\mathcal{U}(\phi)$ coincide with the intersections of the minimal elements of $\mathcal{V}_\chi$ with H, so the next corollary shows why the above example works.

COROLLARY 2. *If B is minimal in $\mathcal{U}(\phi)$, there exist a subgroup B_1 of B and a linear character λ of B_1 such that if μ is the character mentioned in (2), then*
 (i) $\lambda^B = \mu$,
 (ii) $B = \langle B_1^x \mid x \in \Sigma_B \rangle$.

II. **Clifford Theory.** The results of §I are merely a simple application of an elegant new result of Dade [7, (0.4), p. 3] on extensions of stable irreducible characters; so for practical as well as aesthetic reasons, one becomes interested in the extent to which Clifford theory can be done for indecomposable modules. We begin by discussing analogues of the classical theory of extensions of stable modules.

Let H be a normal subgroup of the finite group G, let $X = G/H$, and let M be a G-stable $\mathfrak{D}(G)$-module. As a consequence of Conlon's analysis in [4] it is clear that the centralizing ring

$$\mathcal{C} = \mathrm{Hom}_{\mathfrak{D}(G)}(M^G, M^G)$$

can be written in the form

$$\mathcal{C} = \sum_{x \in X} \oplus \, \mathcal{C}_x$$

where each summand $\mathcal{C}_x$ is an additive abelian subgroup of $\mathcal{C}$, and for $x, y \in X$,

$$\mathcal{C}_x \mathcal{C}_y = \mathcal{C}_{xy} \qquad (\mathfrak{D}\text{-module product}).$$

Consequently $\mathcal{C}_1$ is an $\mathfrak{D}$-subalgebra of $\mathcal{C}$. Further $\mathcal{C}_1$ contains the identity of $\mathcal{C}$ and is isomorphic to $\mathrm{Hom}_{\mathfrak{D}(H)}(M, M)$ as an $\mathfrak{D}$-algebra. In the terminology of [7] we have just noted that $\mathcal{C}\{\mathcal{C}_x \mid x \in X\}$ is a X-graded Clifford system. Let $\mathcal{V}(\mathcal{C}_1)$ denote the group of units of $\mathcal{C}_1$.

THEOREM 3. *Let M be a G-stable $\mathfrak{D}(H)$-module. Then there exists an extension (not necessarily central)*

$$1 \to \mathscr{V}(\mathscr{C}_1) \to G^* \to G \to 1$$

and an abelian group M^* such that

(i) M^* is a right $R(G^*)$-module,
(ii) M^* is a left $\mathscr{C}_1$-module,
(iii) *The abelian group $\mathscr{C} \otimes_{\mathscr{C}_1} M^*$ has a right $\mathfrak{D}(G)$-module structure such that as $\mathfrak{D}(G)$-modules,*

$$\mathscr{C} \otimes_{\mathscr{C}_1} M^* \cong M \otimes_{\mathfrak{D}(H)} \mathfrak{D}(G) = M^G.$$

We remark that if $\sqrt{\mathscr{C}_1}$ denotes the radical of $\mathscr{C}_1$, and if M is an indecomposable $\mathfrak{D}(G)$-module, then our hypothesis implies $\mathscr{C}/\sqrt{\mathscr{C}_1}\mathscr{C}$ is a twisted group algebra for X over the residue class field $\mathfrak{D}/\mathfrak{P}$. The significance of this twisted group algebra for the decomposition of M^G into a direct sum of indecomposable $\mathfrak{D}(G)$-modules was first noted in the paper [**4**] by Conlon and in the series of papers [**12**], [**13**], [**14**] by Tucker. As a corollary of Theorem 3, we can decompose M^G, obtaining a factorization of the indecomposable components of M^G as tensor products.

COROLLARY 4. *Let M be indecomposable. Let $e_1, \ldots, e_n$ denote a complete set of primitive orthogonal idempotents in $\mathscr{C}$, then*

$$M^G \cong \mathscr{C} \otimes_{\mathscr{C}_1} M^* \cong \sum_{i=1}^{n} \oplus e_i \mathscr{C} \otimes_{\mathscr{C}_1} M^*$$

is a decomposition of M^G into a direct sum of indecomposable $\mathfrak{D}(G)$-modules.

It is possible to go further. Not every indecomposable $\mathfrak{D}(G)$-module I for which I_H is isomorphic to a direct sum of copies of M need be a component of M^G, so we can ask if it is possible to factor such modules. Let $\boldsymbol{M}_\mathscr{C}(\mathscr{C}_1\text{-}free)$ denote the category of right $\mathscr{C}$-modules which are free as $\mathscr{C}_1$-modules, and let $\boldsymbol{M}_{\mathfrak{D}(G)}(M\text{-}homogeneous)$ denote the category of right $\mathfrak{D}(G)$-modules I such that I_H is isomorphic to a direct sum of copies of M.

THEOREM 5. *Let M be a G-stable finitely generated $\mathfrak{D}(H)$-module, then $\otimes_{\mathscr{C}_1} M^*$: $\boldsymbol{M}_\mathscr{C}(\mathscr{C}_1\text{-}free) \to \boldsymbol{M}_{\mathfrak{D}(G)}(M\text{-}homogeneous)$ is a full and faithful functor. In more pedestrian terms,*

$$\otimes_{\mathscr{C}_1} M^* : J \to J \otimes_{\mathscr{C}_1} M^* \qquad (J \in \boldsymbol{M}_\mathscr{C}(\mathscr{C}_1\text{-}free))$$

is 1-1 on isomorphism classes, and for any two objects J,K in $\boldsymbol{M}_\mathscr{C}(\mathscr{C}_1\text{-}free)$, the map $\varphi \to \varphi \otimes 1$ defines a canonical isomorphism

$$\operatorname{Hom}_\mathscr{C}(J,K) \cong \operatorname{Hom}_{\mathfrak{D}(G)}(J \otimes_{\mathscr{C}_1} M^*, K \otimes_{\mathscr{C}_1} M^*).$$

This, of course, does not say we can always factor objects in $\boldsymbol{M}_{\mathfrak{D}(G)}(M\text{-homo-}$ geneous) since it does not guarantee $\otimes_{\mathscr{C}_1} M^*$ is surjective on isomorphism classes. In view of Theorem 5, surjectivity on isomorphism classes is equivalent to the question

(3) Is $\otimes_{\mathscr{C}_1} M^* : M_{\mathscr{C}}(\mathscr{C}_1\text{-free}) \to M_{\mathfrak{D}(G)}(M\text{-homogeneous})$
an equivalence of categories?

THEOREM 6. *Let $\mathscr{B}$ denote the centralizer of $\mathscr{C}_1$ in $\mathscr{C}$. Assume for each x in X that $\mathscr{B} \cap \mathscr{C}_x$ contains a unit of $\mathscr{C}$. Then $\otimes_{\mathscr{C}_1} M^* : M_{\mathscr{C}}(\mathscr{C}_1\text{-free}) \to M_{\mathfrak{D}(G)}(M\text{-homogeneous})$ is an equivalence, i.e. $\otimes_{\mathscr{C}_1} M^* : J \to J \otimes_{\mathscr{C}_1} M^*$ defines a bijection on isomorphism classes of objects.*

III. **Vertices and sources.** Up to this point, we have only a generalization of the theory of extensions of stable characters. Dade's work [7], however, goes far beyond this, and we are interested, in particular, in analogues to his theorem [7, (0.4), p. 3] mentioned above. We do not yet have a complete analogue but part of his argument does go through.

Consider the following situation. Let H be normal in G, let M be a G-stable indecomposable $\mathfrak{D}(G)$-module, let B be a vertex of M (in H), and let J denote the (unique up to isomorphism) source of M on $N_H(B)$. The Frattini argument shows that $N_G(B)$ covers G/H. Then as in §II,

$$\mathscr{D} = \mathrm{Hom}_{\mathfrak{D}(N_G(B))}(J^{N_G(B)}, J^{N_G(B)})$$

is an X-graded Clifford system,

$$\mathscr{D} = \sum_{x \in X} \oplus \mathscr{D}_x$$

where $\mathscr{D}_1$ is isomorphic to $\mathrm{Hom}_{\mathfrak{D}(N_H(B))}(J,J)$. If $\mathscr{C}$ is the Clifford system discussed in §II, then under the hypothesis of this paragraph, we have

THEOREM 7. *$\mathscr{C}/\sqrt{\mathscr{C}_1}\mathscr{C}$ is isomorphic to $\mathscr{D}/\sqrt{\mathscr{D}_1}\mathscr{D}$ as twisted group algebras over the residue class field $\mathfrak{D}/\mathfrak{P}$.*

If we change our assumptions on $\mathfrak{D}$, assuming $\mathfrak{D}$ is an algebraically closed field of characteristic $p > 0$, and, in addition, assume that M and J are irreducible, then the results of §§II and III yield

COROLLARY 8. *There is a 1-1 correspondence between isomorphism classes of objects in $M_{\mathfrak{D}(G)}(M\text{-homogeneous})$ and isomorphism classes of objects in $M_{\mathfrak{D}(N_G(B))}$ (J-homogeneous).*

REMARKS. (1) An example due to Tom Berger shows that the correspondence of Corollary 8 is not, in general, the Green correspondence, even though it is canonical.

(2) The results of the last two sections of this note can be generalized and simplified by placing them in the context of Dade's theory of Clifford systems.

Finally, in concluding this note, I would like to thank Professor E. C. Dade for the many helpful comments he has made during the course of this work; in particular, he suggested that results similar to those of §II might hold, and this comment led directly to those results.

References

1. E. Cline, *On minimal vertices and the degrees of irreducible characters* (to appear).

2. ———, *Stable Clifford theory* (to appear).

3. ———, *An application of Dade's method to the theory of vertices and sources* (to appear).

4. S. B. Conlon, *Twisted group algebras and their representations*, J. Austral. Math. Soc. **4** (1964), 152–173. MR **29** #5921.

5. E. C. Dade, *Characters and solvable groups*, University of Illinois, Urbana, Ill., 1967 (preprint).

6. ———, *Compounding Clifford's theory*, Ann. of Math. (submitted).

7. ———, *Isomorphisms of Clifford extensions*, Université de Strasbourg, 1969 (preprint).

8. W. Feit, *Representations of finite groups*, Yale University, New Haven, Conn., 1969 (mimeograph).

9. J. A. Green, *On the indecomposable representations of a finite group*, Math. Z. **70** (1958/59), 430–445. MR **24** #A1304.

10. ———, *Blocks of modular representations*, Math. Z. **79** (1962), 100–115. MR **25** #5114.

11. ———, *A transfer theorem for modular representations*, J. Algebra **1** (1964), 73–84. MR **29** #147.

12. P. A. Tucker, *On the reduction of induced representations of finite groups*, Amer. J. Math. **84** (1962), 400–420. MR **26** #1353.

13. ———, *Note on the reduction of induced representations*, Amer. J. Math. **85** (1963), 53–58. MR **27** #2564.

14. ———, *On the reduction of induced indecomposable representations*, Amer. J. Math. **87** (1965), 798–806. MR **33** #2734.

University of Minnesota

Finite Groups Admitting Almost Fixed-Point-Free Automorphisms

Michael J. Collins

In [1], Thompson proved that a finite group admitting an automorphism of prime order fixing only the identity is nilpotent. Here we shall consider groups satisfying the following hypothesis which, in a sense, is a next stage.

(*) *G is a finite group admitting an automorphism α of order r whose fixed point subgroup P_α has order q, where q and r are distinct primes.*

THEOREM A. *Let G be a group satisfying hypothesis* (*). *Assume either that q is odd or that $q = 2$ and r is not a Fermat prime greater than 3. Then G is soluble.*

The proof of this theorem is by induction on the order of G for each pair q and r. A minimal counterexample is readily seen to be simple, with every proper α-invariant subgroup soluble. Thus a similar problem to that for Thompson's theorem arises. What is the precise structure of such subgroups, and how does one convert information about this limited collection to a wider class of subgroups? In particular, can we control fusion? The answer, fortunately, is yes—for odd Sylow subgroups.

Certain elementary observations and consequences of (*) holding in any such group are crucial:

(1) Thompson's theorem applies to any section admitting α fixed-point-freely;

(2) If $G_\alpha = \langle x \rangle$, then $N(\langle x \rangle) = C(x)$, and there is an α-invariant Sylow q-subgroup Q of G containing x such that $C(x) \subseteq Q \cdot C(Q)$;

(3) If $N(Q)$ is not nilpotent, and $p \mid [N(Q):Q \cdot C(Q)]$, then a Sylow p-subgroup of G has a nilpotent normaliser;

(4) α fixes a Sylow subgroup for each prime divisor of $|G|$.

The structure of soluble groups is given by the following:

AMS 1968 *subject classifications.* Primary 2025.

THEOREM B. *Let G be a soluble group satisfying (*). Then either*
  (i) *Q is normal in G, and hence G/Q is nilpotent; or*
  (ii) *if $G_1 = O_{q,q'}(G)$, then*
      (a) *$x \notin G_1$;*
      (b) *G_1 is nilpotent;*
      (c) *x does not centralise $O_q(G)$; and*
      (d) *G/G_1 is a q-group, cyclic of order q except possibly if $q = 2$ and $r = 2^n + 1$
(in which case it may be extra-special of order 2^{2n+1}).*

It follows that for each prime divisor p of $|G|$, G has either a normal Sylow p-subgroup or a normal p-complement. If (i) holds, but G is not nilpotent, we say that G is of *type I*, and if (ii) holds, that G is of *type II*. It might be mentioned that it is the final possibility in (d) that prevents a proof of solubility in the exceptional cases.

In a minimal counterexample G to Theorem A, we first study the maximal α-invariant subgroups. Each must be a Sylow normaliser. Suppose that M is one, and P is a Sylow p-subgroup of M, such that M has a normal p-complement. Then $M = N(J(P)) = C(Z(P))$, and Glauberman's modification of the Thompson normal p-complement theorem [2], [3] shows that the existence of such an M implies that $p = 2$ and that the symmetric group S_4 is involved in G. Then assertion (3) enables us to make the following deductions:

(5) S_4 is involved in G, and so $r \geq 5$. (Indeed, (*) together with the assumption that S_4 is not involved implies solubility.)

(6) If S is a Sylow 2-subgroup of G, then $N(S)$ is nilpotent.

(7) Let M be a maximal α-invariant subgroup of G. Then
  (i)   if M is nilpotent, M is a Sylow 2-subgroup of G;
  (ii) if M is of type I, then q is odd, $M = N(Q)$, and M is a $\{2,q\}$-group; and
  (iii) if M is of type II, $M = N(H)$ where $H = O_{q'}(M)$ is a Hall subgroup of G of odd order, and if q is odd, $M/O_q(M)$ is a Frobenius group.

At this stage, the cases q odd and $q = 2$ must be considered separately.

If q is odd, it can be shown that Q is elementary abelian: otherwise Glauberman's weakly closed elements theorem [4] can be used to show that $\Phi(Q) \cap Z(Q)$ is weakly closed in Q. Then a maximality argument shows that $N(Q)$ contains a Sylow 2-subgroup of G controlling its own fusion, which is impossible in a simple group. As a consequence, if M is a maximal α-invariant subgroup of type II, $O_q(M) = 1$ so that $M = H\langle x \rangle$ is a Frobenius group. Now, for P a Sylow p-subgroup of H, $M = N(Z(P)) = N(J(P))$ and $H = C(Z(P))$. The following result generalising Glauberman's weakly closed elements theorem enables us to show that $Z(P)$ is weakly closed in P since $[N(Z(P)):C(Z(P))] = q$.

THEOREM. *Let G be a finite group, p a prime, and P a Sylow p-subgroup of G. Let Q be a subgroup of $Z(P)$ which is normal in $N(J(P))$. If either*
  (i) *p is odd, and $(p-1)$ does not divide $[N(Q):C(Q)]$, or*
  (ii) *$Qd(p)$ is not involved in G, then Q is weakly closed in P with respect to G.*

(This can be proved by a direct generalisation of Glauberman's proof. See also Corollary 3 of [5] for an independent proof.)

As a result, G is p-normal, and fusion in P is controlled in $N(Z(P))$, i.e. in M. This is actually an elementary result if H is not of prime power order, but requires the above theorem if it is. Under these circumstances one can expect the p-local subgroups of G to inherit "reasonable" properties from $N(Z(P))$. What turns out to be relevant here is that if P_1 is a nonidentity p-subgroup of G, then (i) $C(P_1)$ has a normal p-complement, and (ii) $N(P_1)/C(P_1)$ has a normal p-subgroup of index 1 or q. The same holds in sections. Thus if $q \neq 3$, no section can be isomorphic to S_4, contrary to (5). Hence Theorem A is proved for $q \geq 5$.

If $q = 3$, a more detailed study of the local structure is necessary. It can be shown that Q is a T.I.-subgroup, and by playing off elements of order 3 against involutions that centralise them, it can be shown that Q is self-centralising. Also, if V is a four-group such that $3 \mid |C(V)|$, then $C(V)$ has a normal Sylow 3-subgroup. This is sufficient to show that the symmetric group S_4 cannot be "badly located" in G, and that there are involutions for which Glauberman's weakly closed elements theorem does not "break down." Thus Glauberman's Z^*-theorem [6] yields a contradiction.

If $q = 2$ and r is not a Fermat prime, an elementary transfer argument shows that $C(x)$ is a 2-group. Thus if M is a maximal α-invariant subgroup of type II and $H = O_{2'}(M)$, H is abelian, each element being inverted by x. Furthermore, if $g \in H^\#$, then $C(g) = H \cdot O_2(C(g))$. Under these circumstances, character theory can be employed to show that G has exactly two nonconjugate such subgroups, and that the corresponding Hall subgroups of odd order have orders differing by 2. Since each has order congruent to 1 modulo r, and r is odd, this is impossible, so that the proof of Theorem A is complete.

The treatment of this case involves a special case of the following result [7]. For a group G and subgroup H, define $a(H) = [N_G(H):C_G(H)]$.

THEOREM. *Let G be a finite simple group such that whenever H is a subgroup of odd order which is either abelian or of prime power order, then $a(H) \leq 3$. Then G is isomorphic to one of the groups* $\mathrm{SL}(2,2^a)$, $a \geq 2$, *or* $\mathrm{PSL}(2,7)$.

REFERENCES

1. J. G. Thompson, *Finite groups with fixed-point-free automorphisms of prime order*, Proc. Nat. Acad. Sci. U.S.A. **45** (1959), 578–581. MR **21** #3484.

2. ———, *Normal p-complements for finite groups*, J. Algebra **1** (1964), 43–46. MR **29** #4793.

3. G. Glauberman, *Subgroups of finite groups*, Bull. Amer. Math. Soc. **73** (1967), 1–12. MR **35** #4294.

4. ———, *Weakly closed elements of Sylow subgroups*, Math. Z. **107** (1968), 1–20.

5. ———, *A sufficient condition for p-stability*, J. Algebra (to appear).

6. ———, *Central elements in core-free groups*, J. Algebra **4** (1966), 403–420. MR **34** #2681.

7. M. J. Collins, *Finite groups having subgroups of odd order with small automisers*, Illinois J. Math. (to appear).

UNIVERSITY OF ILLINOIS AT CHICAGO CIRCLE AND
UNIVERSITY COLLEGE, OXFORD

Some Remarks on the Krull-Schmidt Theorem

S. B. Conlon

Let R be a ring and $\mathscr{M}$ the category of left R-modules. Let $\mathscr{C}$ be a class of such modules, closed under isomorphism and consisting only of modules with local endomorphism rings. A module M is said to have a $\mathscr{C}$-*decomposition* if it can be written

(1) $$M = (\bigoplus_{f \in F} M_f) \oplus M^F,$$

with each $M_f \in \mathscr{C}$ and where M^F contains *no* direct summands out of $\mathscr{C}$. Let

(2) $$M = (\bigoplus_{g \in G} N_g) \oplus N^G$$

be a second such $\mathscr{C}$-decomposition of M. For $U \in \mathscr{C}$, let $m_U(n_U)$ denote the cardinal number of summands $M_f(N_g)$ with $M_f \approx U$ ($N_g \approx U$). In (1) (in (2)) put $M_F = \bigoplus_{f \in F} M_f$ ($N_G = \bigoplus_{g \in G} N_g$) and let p_F, $p^F(q_G, q^G)$ denote the projection endomorphisms in (1) (in (2)).

THEOREM. $m_U = n_U$ *for all* $U \in \mathscr{C}$. *Also* $M_F \cap N^G = N_G \cap M^F = (0)$ *and* q_G, q^G *carry* M_F, M^F *monomorphically into* N_G, N^G *respectively.*

REMARK. 1. If $M^F = (0)$, then $N^G = (0)$ and so the usual Krull-Schmidt (-Azumaya) theorem applies.

2. If F is finite, so is G and $M^F \approx N^G$ (and of course $M_F \approx N_G$).

3. If M has a decomposition into indecomposables, we get a $\mathscr{C}$-decomposition by gathering into M_F all indecomposable summands in $\mathscr{C}$ and the remaining indecomposable summands into M^F.

A direct elementary proof of this is given by the author in [1]. The proof is based

---

AMS 1970 subject classifications. Primary 16A48.

on Azumaya's lemma. In this paper $\mathscr{C}$ was in particular taken to be the class of all modules with local endomorphism rings, and little modification is necessary.

However the techniques of Crawley and Jónsson [2] readily lead to a proof. We say a module M_1 has the exchange property if whenever we have

$$M = M_1 \oplus M_2 = \bigoplus_{f \in F} M_f,$$

then for each $f \in F$ we have $M_f = M_f^1 \oplus M_f^2$ and

$$M = M_1 \oplus (\bigoplus_{f \in F} M_f^2).$$

Warfield [3] has noted that an indecomposable module has the exchange property if and only if it has a local endomorphism ring. From Crawley and Jónsson we have that a finite sum of modules each of which has the exchange property has itself the exchange property.

Suppose then that we have a $\mathscr{C}$-decomposition and any other decomposition of M:

$$M = (\bigoplus_{f \in F} M_f) \oplus M^F = (\bigoplus_{g \in G} M_g).$$

Take $F' = (f_1, \ldots, f_n) \subseteq F$. Then we may write

(3) $$N_g = N_g^1 \oplus N_g^2 \qquad (g \in G)$$

and

$$M = (\bigoplus_{f \in F'} M_f) \oplus (\bigoplus_{g \in G} N_g^2).$$

(Only a finite number of $N_g^1 \neq (0)$.)

In proving the above theorem the $\bigoplus N_g$ would be a $\mathscr{C}$-decomposition. The above decomposition (3) is also obtainable from Azumaya's lemma and is the principal step in proving the cardinal identity of $\mathscr{C}$-decompositions.

However the notion of "exchange property" or similar ones such as "refinable" have been introduced to investigate the refinement problem: when do decompositions have isomorphic refinements?

If F is finite in a $\mathscr{C}$-decomposition, the last lemma (3) shows us that any direct summand of M also has a $\mathscr{C}$-decomposition and so any decomposition may be refined to a $\mathscr{C}$-decomposition. The author has been unable to make any progress when F is infinite. Most such theorems generally involve only a countable number of summands where one can apply some kind of diagonal argument or again the summands themselves are countably generated so that this latter case can be reduced to the countable number of summands case. The reduction argument fails and also the countable number of summands case presents difficulties because of a continual throw-back into the factor M^F in which there is no control as to exchange property or number of generators.

If R is a local ring and if G is a finite group, then in considering finitely generated R-free RG-modules, we have the Krull-Schmidt theorem in its strongest form. We are assured of a decomposition into a finite number of indecomposables and so there exist $\mathscr{C}$-decompositions with F finite. Thus the refinement theorem holds for $\mathscr{C}$-decompositions. We also have the cancellation for modules in $\mathscr{C}$, i.e. $M_1 \oplus X \approx M_1 \oplus Y$ with $M_1 \in \mathscr{C}$, implies $X \approx Y$.

References

1. S. B. Conlon, *An extension of the Krull-Schmidt theorem*, Bull. Austral. Math. Soc. **1** (1969), 109–114. MR **40** #1422.

2. Peter Crawley and Bjarni Jónsson, *Refinements for infinite direct decompositions of algebraic systems*, Pacific J. Math. **14** (1964), 797–855. MR **30** #49.

3. R. B. Warfield, Jr., *A Krull-Schmidt theorem for infinite sums of modules*, Proc. Amer. Math. Soc. **22** (1969), 460–465. MR **39** #4213.

UNIVERSITY OF SYDNEY, AUSTRALIA

A Clifford Theory for Blocks

E. C. Dade

Everybody knows about the theory of Clifford [2] relating the characters of a finite group H to those of a normal subgroup K of H and to projective characters of certain subgroups of $G = H/K$. The factor group G acts by conjugation on the family of all irreducible $\mathfrak{F}$-characters of K, where $\mathfrak{F}$ is an algebraically closed field. Let G_φ be the subgroup of G fixing such a character φ. Then there is a well-defined Clifford extension $H\langle\varphi\rangle$, having the multiplicative group F of $\mathfrak{F}$ as a central subgroup and G_φ as factor group $H\langle\varphi\rangle/F$. Furthermore Clifford's theory gives a natural one-to-one correspondence between all irreducible $\mathfrak{F}$-characters of the twisted group algebra $\mathfrak{F}[H\langle\varphi\rangle]$ of G_φ corresponding to the extension $H\langle\varphi\rangle$, and all irreducible $\mathfrak{F}$-characters of H having φ as a K-constituent (see [3] for details).

We want to develop a similar theory for blocks. This turns out to be quite possible, although there are some additional complications due to certain conjugation actions which have no counterpart in the simpler theory for characters.

Since I prefer to keep all fields algebraically closed, the ground ring $\mathfrak{R}$ will be any valuation ring whose field of fractions $\mathfrak{F}$ is algebraically closed. Then the residue class field $\bar{\mathfrak{F}} = \mathfrak{R}/\mathfrak{p}$ of $\mathfrak{R}$ modulo its maximal ideal $\mathfrak{p}$ is also algebraically closed. Notice that it is quite possible to have $\mathfrak{p} = 0$ and $\mathfrak{F} = \mathfrak{R} = \bar{\mathfrak{F}}$. So we can talk about rings and fields simultaneously. In any case, orders $\mathfrak{D}$ over $\mathfrak{R}$ (i.e., associative $\mathfrak{R}$-algebras with identity which are finitely-generated free $\mathfrak{R}$-modules) behave properly. For example, the residue algebra $\bar{\mathfrak{D}} = \mathfrak{D}/\mathfrak{p}\mathfrak{D}$ is a finite-dimensional algebra over $\bar{\mathfrak{F}}$, and the Jacobson radical $J(\mathfrak{D})$ is the inverse image of that of $\bar{\mathfrak{D}}$. Furthermore, every idempotent of $\bar{\mathfrak{D}}$ is the image of at least one idempotent of $\mathfrak{D}$. This implies that the images $\bar{e}_1, \ldots, \bar{e}_n$ in $\bar{\mathfrak{D}}$ of the primitive central idempotents $e_1, \ldots, e_n$ of $\mathfrak{D}$ are precisely the primitive central idempotents of $\bar{\mathfrak{D}}$. So the blocks

---

AMS 1970 *subject classifications*. Primary 20C15; Secondary 20F25.

of $\mathfrak{D}$ are in one-to-one correspondence with those of $\bar{\mathfrak{D}}$.

The $\mathfrak{R}$-order we are interested in is, of course, the group ring $\mathfrak{D} = \mathfrak{R}H$ of H over $\mathfrak{R}$. For any $\sigma \in G = H/K$, let $\mathfrak{D}_\sigma$ be the $\mathfrak{R}$-submodule of $\mathfrak{D}$ having the elements of the coset σ as basis. Then we have:

(1a) $$\mathfrak{D} = \bigoplus \sum_{\sigma \in G} \mathfrak{D}_\sigma \quad (as\ \mathfrak{R}\text{-}modules),$$

(1b) $$\mathfrak{D}_\sigma \mathfrak{D}_\tau = \mathfrak{D}_{\sigma\tau} \quad (module\ product),\ for\ all\ \sigma, \tau \in G.$$

The theory we shall describe is valid for any $\mathfrak{R}$-order $\mathfrak{D}$ having a decomposition of the above type.

Let $\mathfrak{C}$ be the centralizer of the subring $\mathfrak{D}_1 = \mathfrak{R}K$ in $\mathfrak{D}$, and $\mathfrak{C}_\sigma$ be the centralizer of $\mathfrak{D}_1$ in $\mathfrak{D}_\sigma$, for any $\sigma \in G$. Then $\mathfrak{C}$ is a suborder of $\mathfrak{D}$, and the $\mathfrak{C}_\sigma$ are $\mathfrak{R}$-submodules of $\mathfrak{C}$ satisfying:

(2a) $$\mathfrak{C} = \bigoplus \sum_{\sigma \in G} \mathfrak{C}_\sigma \quad (as\ \mathfrak{R}\text{-}modules),$$

(2b) $$\mathfrak{C}_\sigma \mathfrak{C}_\tau \subseteq \mathfrak{C}_{\sigma\tau} \quad (module\ product),\ for\ all\ \sigma, \tau \in G.$$

The change from (1b) to (2b) makes all the difference in the world, and our constructions are motivated by the overwhelming necessity to get back to (1b) for suitable subrings of $\mathfrak{C}$.

Evidently $\mathfrak{C}_1$ is the center of $\mathfrak{D}_1 = \mathfrak{R}K$. So a block B of $\mathfrak{R}K$ is determined by a primitive idempotent e of $\mathfrak{C}_1$. Because $\mathfrak{C}_1$ is central in $\mathfrak{C}$, the product $e\mathfrak{C} = \mathfrak{C}e$ is a suborder of $\mathfrak{C}$. Furthermore, $e\mathfrak{C}_\sigma = \mathfrak{C}_\sigma e$ is an $\mathfrak{R}$-submodule of $\mathfrak{C}_\sigma$, for all $\sigma \in G$, satisfying:

(3a) $$e\mathfrak{C} = \bigoplus \sum_{\sigma \in G} e\mathfrak{C}_\sigma \quad (as\ \mathfrak{R}\text{-}modules),$$

(3b) $$(e\mathfrak{C}_\sigma)(e\mathfrak{C}_\tau) \subseteq e\mathfrak{C}_{\sigma\tau} \quad (module\ product),\ for\ all\ \sigma, \tau \in G.$$

We have not yet got back to (1b). But we define:

(4a) $$G[B] = \{\sigma \in G | (e\mathfrak{C}_\sigma)(e\mathfrak{C}_{\sigma^{-1}}) = e\mathfrak{C}_1\},$$

(4b) $$\mathfrak{C}[B] = \bigoplus \sum_{\sigma \in G[B]} e\mathfrak{C}_\sigma,$$

(4c) $$\mathfrak{C}[B]_\sigma = e\mathfrak{C}_\sigma,\ for\ all\ \sigma \in G[B].$$

Then one can show that $G[B]$ is a normal subgroup of the subgroup G_B of G fixing the block B (i.e., the idempotent e) under conjugation. Furthermore, we now have reached a good system, in that $\mathfrak{C}[B]$ is an $\mathfrak{R}$-order and:

(5a) $$\mathfrak{C}[B] = \bigoplus \sum_{\sigma \in G[B]} \mathfrak{C}[B]_\sigma \quad (as\ \mathfrak{R}\text{-}modules),$$

(5b) $$\mathfrak{C}[B]_\sigma \mathfrak{C}[B]_\tau = \mathfrak{C}[B]_{\sigma\tau} \quad (module\ product),\ for\ all\ \sigma, \tau \in G[B].$$

We still do not have a Clifford extension. To define one, we notice that the radical $J(\mathfrak{C}[B]_1)$ satisfies $\mathfrak{C}[B]_1/J(\mathfrak{C}[B]_1) \cong \bar{\mathfrak{F}}$, since e is a primitive idempotent of the commutative $\mathfrak{R}$-order $\mathfrak{C}_1$. It follows that there is a unique central extension $G[B]^*$ of the multiplicative group $\bar{F}$ of $\bar{\mathfrak{F}}$ by $G[B]$ whose twisted group algebra $\bar{\mathfrak{F}}[G[B]^*]$ is the quotient ring:

$$\bar{\mathfrak{F}}[G[B]^*] = \mathfrak{C}[B]/\mathfrak{C}[B]J(\mathfrak{C}[B]_1), \tag{6}$$

by the two-sided ideal $\mathfrak{C}[B]J(\mathfrak{C}[B]_1) = J(\mathfrak{C}[B]_1)\mathfrak{C}[B]$. The image of $\mathfrak{C}[B]_\sigma \bar{\mathfrak{F}}[G[B]^*]$ is precisely the one dimensional subspace $\bar{\mathfrak{F}}[G[B]^*]_\sigma$ spanned by the inverse image of σ in $G[B]^*$, for any element $\sigma \in G[B]$. Of course, we call $G[B]^*$ the *Clifford extension* for the block B.

The group G operates by conjugation on the suborder $\mathfrak{C}$. Evidently the centralizer $C(G \text{ in } \mathfrak{C})$ of G in $\mathfrak{C}$ is the center of $\mathfrak{D}$. It follows that the primitive central idempotents f of $\mathfrak{D}$ are simply sums of G-conjugacy classes of primitive central idempotents of $\mathfrak{C}$. Using the fact that e is a central idempotent of $\mathfrak{D}$, one easily sees that these idempotents f satisfying $ef \neq 0$ (i.e., those corresponding to blocks $\tilde{B}$ of $\mathfrak{D}$ lying over the block B of $\mathfrak{D}_1$) correspond one-to-one to the sums g of G_B-conjugacy classes of primitive central idempotents of $e\mathfrak{C}$, the correspondence sending f into ef and g into $\text{tr}_{G_B \to G}(g)$. This correspondence was noticed long ago by Reynolds [6].

One can show that any $e\mathfrak{C}_\sigma$, for $\sigma \in G - G[B]$, lies in the radical $J(e\mathfrak{C})$ of $e\mathfrak{C}$. This implies that the central idempotents of $e\mathfrak{C}$ all lie in the suborder $\mathfrak{C}[B]$. From this observation, we conclude easily that the sums g of G_B-conjugacy classes of primitive central idempotents of $e\mathfrak{C}$ are just the similar sums for $\mathfrak{C}[B]$.

Finally, using the fact that $\mathfrak{C}[B]_1 = e\mathfrak{C}_1$ is central in $\mathfrak{C}[B]$, one can prove that the primitive central idempotents of $\mathfrak{C}[B]$ correspond one-to-one with their images in $\bar{\mathfrak{F}}[G[B]^*]$, which are just the primitive central idempotents of that algebra, and hence correspond to the "blocks of the extension $G[B]^*$." The result of all this is:

THEOREM 7. *There is a natural one-to-one correspondence between the blocks $\tilde{B}$ of $\mathfrak{D}$ lying over the block B of $\mathfrak{D}_1$ and the G_B-conjugacy classes of blocks B^* of the Clifford extension $G[B]^*$.*

As remarked above, the preceding theorem is valid in the general situation (1). If we take advantage of the additional structure afforded by our group algebra $\mathfrak{D} = \mathfrak{R}H$ we can say quite a bit more about the Clifford extension $G[B]^*$ and the blocks $\tilde{B}$ and B^*.

We assume now that our residue classes field $\bar{\mathfrak{F}}$ has prime characteristic p. We know from Brauer [1] that the block B of $\mathfrak{D}_1 = \mathfrak{R}K$ determines a unique K-conjugacy class of p-subgroups D of K, its defect groups. Furthermore, given such a defect group D, the block B corresponds to a unique $N_K(D)$-conjugacy class of irreducible $\bar{\mathfrak{F}}$-characters $\varphi_1, \ldots, \varphi_n$ of $C_K(D)$, characters lying in blocks of defect zero of $C_K(D)/Z(D)$. Evidently the group $C_K(D)$ is normal in $C_H(D)$. So the irreducible $\bar{\mathfrak{F}}$-character $\varphi = \varphi_1$ has a Clifford extension $C_H(D)\langle\varphi\rangle$. The subgroup $N_H(D)_\varphi$ of $N_H(D)$ fixing φ acts by conjugation on this Clifford extension. Its normal subgroup

$N_K(D)_\varphi$ centralizes $C_H(D)\langle\varphi\rangle/\bar{F} = C_H(D)_\varphi/C_K(D)$. It follows that there is a "bilinear" map $\omega: N_K(D)_\varphi/DC_K(D) \times C_H(D)_\varphi/C_K(D) \to \bar{F}$ so that:

(8) $\qquad \rho^\sigma = \omega(\sigma DC_K(D), \rho\bar{F})\rho, \quad \text{for all } \rho \in C_H(D)\langle\varphi\rangle, \sigma \in N_K(D)_\varphi.$

We denote by $C_H(D)_\omega$ the normal subgroup of all $\tau \in C_H(D)_\varphi$ satisfying $\omega(N_K(D)_\varphi/DC_K(D), \tau C_K(D)) = \{1\}$, and by $C_H(D)_\omega\langle\varphi\rangle$ the inverse image in $C_H(D)\langle\varphi\rangle$ of $C_H(D)_\omega/C_K(D)$. Then $N_H(D)_\varphi/N_K(D)_\varphi$ acts naturally by conjugation on the central extension $C_H(D)_\omega\langle\varphi\rangle$ of $\bar{F}$ by $C_H(D)_\omega/C_K(D)$.

With all this notation, we can now compute $G[B]$, $G[B]^*$, and the action of G_B on $G[B]^*$.

THEOREM 9. *The group $G[B]$ is $C_H(D)_\omega K/K$. The inclusion isomorphism of $C_H(D)_\omega/C_K(D)$ onto $G[B]$ carries the extension $C_H(D)_\omega\langle\varphi\rangle$ onto the Clifford extension $G[B]^*$. The group G_B is $N_H(D)_\varphi K/K$. The inclusion isomorphism of $N_H(D)/N_K(D)_\varphi$ onto G_B together with the above isomorphism of $C_H(D)_\omega\langle\varphi\rangle$ onto $G[B]^*$ carry the action of $N_H(D)_\varphi/N_K(D)_\varphi$ on $C_H(D)_\omega\langle\varphi\rangle$ onto that of G_B on $G[B]^*$.*

We can also compute the defect groups of a block $\tilde{B}$ of $\Re H$ lying over B in terms of certain defect groups of B and of a corresponding block B^* of $G[B]^*$. The conjugacy action of the inverse image H_B (of G_B in H) on the ring $\Re K$ defines, as in Green [5], a family of defect groups D_H of B in H_B. One easily sees that $D_H \cap K$ is an ordinary defect group D of B, while $D_H K/K$ is a p-Sylow subgroup of $G_B = H_B/K$. By conjugacy, we can suppose that $D_H K/K$ contains a defect group D^* of the block B^* in $(G_B)_{B^*}$ (defined by the conjugation action of that group on $\bar{\mathfrak{F}}[G[B]^*]$). Then we can prove:

THEOREM 10. *The inverse image $\tilde{D}$ of D^* in D_H is a defect group of $\tilde{B}$ in H.*

Finally, we should note that $G[B]$ fixes every irreducible character Ψ in the block B. If Ψ is an $\bar{\mathfrak{F}}$-character, then the Clifford extension $G[B]\langle\Psi\rangle$ is naturally isomorphic to $G[B]^*$. If Ψ is an $\mathfrak{F}$-character, then $G[B]^*$ is naturally isomorphic to the "$\Re/\mathfrak{p}$-residual" (as in §4 of [4]) of the Clifford extension $G[B]\langle\Psi\rangle$.

BIBLIOGRAPHY

1. R. Brauer, *Zur Darstellungstheorie der Gruppen von endlicher Ordnung. I*, Math. Z. **63** (1956), 406–444. MR **17**, 824.
2. A. H. Clifford, *Representations induced in an invariant subgroup*, Ann. of Math. (2) **38** (1937), 533–550.
3. E. C. Dade, *Compounding Clifford's theory*, Ann. of Math. (2) **91** (1970), 236–290.
4. ———, *Isomorphisms of Clifford extensions*, Ann. of Math. (to appear).
5. J. A. Green, *Some remarks on defect groups*, Math. Z. **107** (1968), 133–150. MR **38** #2222.
6. W. F. Reynolds, *Blocks and normal subgroups of finite groups*, Nagoya Math. J. **22** (1963), 15–32. MR **27** #3690.

INSTITUTE DE RECHERCHE MATHÉMATIQUE AVANCÉE DE STRASBOURG

Jordan's Theorem for Solvable Groups

Larry Dornhoff

THEOREM. *Let n be a positive integer, C the field of complex numbers, G a finite solvable subgroup of $\mathrm{GL}(n,C)$. Then there is an abelian normal subgroup A of G such that*

$$|G:A| \leq 2^{4n/3-1} 3^{10n/9-1/3}.$$

When $n = 3 \cdot 4^k$, $k = 0, 1, \ldots$, this bound is attained.

This theorem has already appeared in [3]. We take this opportunity to shorten the proof and make it independent of the main results in [1], [2].

Consider the case when G is irreducible and primitive as a linear group. If $\mathrm{Fit}(G)$ is the Fitting subgroup of G, write

$$\mathrm{Fit}(G) = P_1 \times \cdots \times P_k \times P_{k+1} \times \cdots \times P_l,$$

P_i a Sylow p_i-subgroup of $\mathrm{Fit}(G)$, the notation chosen so that $P_1, \ldots, P_k$ are nonabelian and $P_{k+1}, \ldots, P_l$ are abelian. Since G is primitive, Clifford's theorem implies that abelian normal subgroups of G are cyclic and central. Hence $P_{k+1}, \ldots, P_l \subseteq Z(G)$. If $i \leq k$, characteristic abelian subgroups of P_i are cyclic and we use Philip Hall's theorem on the structure of such p_i-groups. We conclude that

$$P_i = E_i \circ Z_i \quad \text{(central product)},$$

E_i extra-special, $Z_i \subseteq Z(G)$ cyclic. If $|E_i| = p_i^{2m_i+1}$, then faithful irreducible representations of P_i all have degree $p_i^{m_i}$, and we use Clifford's theorem for $P_i \triangleleft G$ to conclude $p_i^{m_i} | n$. Thus we have proved $|\mathrm{Fit}(G):Z(G)| | n^2$, a result of Suprunenko.

AMS 1970 *subject classifications*. Primary 20C15, 20H20.

The automorphism group of E_i is known, and we quickly see that

(*) $\quad\quad\quad$ $G/\text{Fit}(G)$ is isomorphic to a solvable subgroup
of $\text{Sp}(2m_1, p_1) \times \cdots \times \text{Sp}(2m_k, p_k)$.

It is now enough to show $|G:\text{Fit}(G)| \leq n^{-2} 2^{4n/3-1} 3^{10n/9-1/3}$, taking $Z(G)$ as our A. $|\text{Sp}(2m_i, p_i)| < p_i^{m_i(2m_i+1)}$, so we have

$$|G:\text{Fit}(G)| < \prod_i p_i^{m_i(2m_i+1)} \leq \prod_i p_i^{m_i(2 \text{ Max}\{m_i\}+1)}$$
$$\leq n^{2 \text{ Max}\{m_i\}+1} \leq n^{2 \log_2 n + 1}.$$

This proves the result for $n \geq 13$. For $n \leq 12$, (*) gives the result except when $n = 4$ or $n = 8$. If $n = 4 = 2^2$ or $n = 8 = 2^3$, we use the following two easy facts to complete the proof:

$\quad\quad\quad$ $\text{Sp}(4,2)$ has no solvable subgroup of index ≤ 6.

$\quad\quad\quad$ $\text{Sp}(6,2)$ has no proper subgroup of index ≤ 9.

The imprimitive and reducible cases for G follow as in [3]. We finally remark that the group G^* of [3] is easy to construct. Use a faithful representation to imbed the extra-special group P_0 of order 27 and exponent 3 in $\text{SL}(3,C)$, and then choose $G^* = N_{\text{SL}(3,C)}(P_0)$.

References

1. H. F. Blichfeldt, *Finite collineation groups*, Univ. of Chicago Press, Chicago, Ill., 1917.

2. John D. Dixon, *The Fitting subgroup of a linear solvable group*, J. Austral. Math. Soc. **7** (1967), 417–424. MR **37** #6372.

3. Larry Dornhoff, *Jordan's theorem for solvable groups*, Proc. Amer. Math. Soc. **24** (1970), 533–537.

UNIVERSITY OF ILLINOIS, URBANA

Operations in Representation Rings

Andreas Dress

0. Introduction. In the following lecture I will first define a certain class of not necessarily additive maps from an abelian semigroup A into an abelian group B, which factor uniquely through the universal "Grothendieck"-group $\bar{A}$ of A. In the following parts I want to apply this concept to certain situations which occur in representation theory of finite groups.

1. Algebraic maps.

DEFINITIONS. Let A be an abelian semigroup and B an abelian group. A set-theoretic map $\varphi : A \to B$ is called algebraic of degree 0 if φ is constant. It is called algebraic of degree $\leq n + 1$ if for every $x \in A$ the map $D_x\varphi : A \to B : a \to \varphi(x + a) - \varphi(a)$ is algebraic of degree $\leq n$.

It is called algebraic if there is some n such that φ is algebraic of degree $\leq n$. Moreover, we define degree $\varphi = \text{Min}\{n | \varphi \text{ algebraic of degree } \leq n\}$.

EXAMPLES. 1. Any homomorphism $\varphi : A \to B$ is algebraic of degree ≤ 1. More generally a map $\varphi : A \to B$ is algebraic of degree ≤ 1 if and only if φ is the sum of a homomorphism $\psi : A \to B$ and a constant $b \in B : \varphi(a) = \psi(a) + b$.

2. A map $\varphi : \mathbf{Q} \to \mathbf{Q}$ is algebraic if and only if there exists some polynomial $f(x) \in \mathbf{Q}[x]$ with $f(q) = \varphi(q)$ for all $q \in \mathbf{Q}$. Moreover, degree $f = $ degree φ.

3. If $\text{Hom}(A,B) = 0$, then any algebraic map $\varphi : A \to B$ is constant.

4. For elementary abelian p-groups A and B every set-theoretic map $A \to B$ is algebraic.

We have the following two propositions:

PROPOSITION 1.1. *If $\varphi : A \to B$ is algebraic and if $\iota : A \to \bar{A}$ is the canonical map*

AMS 1970 *subject classifications*. Primary 20C15, 20C05; Secondary 20D10, 20D15, 20J05.

from A into its associated universal group $\bar{A}$, then there exists exactly one algebraic map $\bar{\varphi}:\bar{A} \to B$ such that $\varphi = \bar{\varphi}\iota$:

$$\begin{array}{ccc} A & \xrightarrow{\varphi} & B \\ \iota & \searrow & \nearrow \bar{\varphi} \\ & \bar{A} & \end{array}$$

commutative. Moreover, degree φ = degree $\bar{\varphi}$.

PROPOSITION 1.2. *Let A,B,C be abelian groups and $\varphi:A \to B$, $\psi:B \to C$ algebraic maps. Then $\psi\varphi:A \to C$ is algebraic and degree $\psi\varphi \leq$ degree $\varphi \cdot$ degree ψ.*

The proof of Proposition 1.1 is straightforward, the proof of Proposition 1.2 a bit tricky. They will appear in [3].

REMARK. The formulas occurring in the proof of Proposition 1.2 can be used also to give a general formula for the partial derivatives $\partial g \circ f / \partial x_1 \cdots \partial x_n$ of the composition of highly differentiable maps $f:E_1 \to E_2$, $g:E_2 \to E_3$ between, say, Banach spaces E_1, E_2, E_3 with respect to certain elements $x_1, \ldots, x_n \in E_1$, which are generally proved by other methods from combinatorial analysis (see [5]).

2. **Exponentials in $\Omega(G)$.** Let G be a finite group. A G-set S is a finite set, on which G acts by permutations from the left. The isomorphism classes of G-sets form a "half-ring" $\Omega^+(G)$, if we define: $[S] + [T] = [S \cup T]$, $[S] \cdot [T] = [S \times T]$, where $[S]$ is the isomorphism class of S, $S \cup T$ the disjoint union and $S \times T$ the cartesian product of S and T. Let $\Omega(G)$ be the associated Grothendieck ring, which has also been called the Burnside ring of G, see [1], [2], [6]. We can define an additional structure on $\Omega(G)$ if we consider for any two G-sets S and T the set $S^T = \{f:T \to S | f$ set-theoretic map$\}$ which is a G-set as well with $gf:T \to S$: $t \mapsto g(f(\bar{g}^1 t))$.

We have

PROPOSITION 2.1. (a) $(S_1 \times S_2)^T \cong S_1^T \times S_2^T$,
(b) $S^{T_1 \times T_2} \cong (S^{T_1})^{T_2}$,
(c) $S^{T_1 \cup T_2} \cong S^{T_1} \times S^{T_2}$,
(d) $S^* \cong S$ (* *the one-point trivial G-set*).

PROPOSITION 2.2. *Let T be a fixed G-set. Then the map $\Omega^+(G) \to \Omega(G):[S] \mapsto [S^T]$ is algebraic of degree $|T|$ (number of elements in T).*

Thus by Proposition 1.1 we have a unique algebraic map $\Omega(G) \to \Omega(G): x \to x^T$, which extends the map considered in Proposition 2.2, and which, by the uniqueness and Proposition 1.2, has properties analogous to those stated in Proposition 2.1, i.e. it makes the multiplicative semigroup of $\Omega(G)$ to an $\Omega^+(G)$-semimodule.

Grothendieck-constructions, applied to various sub- and quotient-semimodules of $(\Omega(G), \cdot)$ will then lead to various $\Omega(G)$-modules. We will give one example.

3. **Units in $\Omega(G)$.** From [2] one gets easily:

PROPOSITION 3.1. *Let $\mathfrak{f}(G)$ be the group of units in $\Omega(G)$. Then $\mathfrak{f}(G)$ is a finite, elementary abelian 2-group.*

PROPOSITION 3.2. *If $\mathfrak{f}(G) = \{\pm 1\}$, then G is solvable.*

PROOF. By [2] G is solvable if and only if $\Omega(G)$ is not isomorphic to a product of two rings $R_1 \times R_2$. But $\Omega(G) \cong R_1 \times R_2$ and the torsion freeness of $\Omega(G)$ would imply $\mathfrak{f}(G) = \mathfrak{f}(R_1) \times \mathfrak{f}(R_2)$ and $\pm 1 \in \mathfrak{f}(R_1)$, $\pm 1 \in \mathfrak{f}(R_2)$; thus $|\mathfrak{f}(G)| \geq 4$.

PROPOSITION 3.3. *If G is solvable of odd order, then $\mathfrak{f}(G) = \pm 1$.*

This follows from the analysis of prime ideals in $\Omega(G)$ given in [2].

Thus one is tempted to try to find a direct proof for the following version of the Feit-Thompson Theorem:

If G is of odd order, then $\mathfrak{f}(G) = \pm 1$.

One hint of how to find such a proof can be taken from §2. $\mathfrak{f}(G)$ is the biggest subgroup of the semigroup $(\Omega(G), \cdot)$; thus it is an $\Omega^+(G)$-subsemimodule and, because it is a group, already an $\Omega^+(G)$-module. Thus the action of $\Omega^+(G)$ on $\mathfrak{f}(G)$ extends to an action of $\Omega(G)$ on $\mathfrak{f}(G)$ and makes $\mathfrak{f}(G)$ to an $\Omega(G)$-module. One can prove in a very elementary way:

PROPOSITION 3.4. *The $\Omega(G)$-submodule of $\mathfrak{f}(G)$, generated by $-1 \in \mathfrak{f}(G)$ is $\pm 1 \in \mathfrak{f}(G)$ if and only if G is of odd order.*

Thus on the one hand the two statements in Proposition 3.3 are indeed equivalent, i.e. $\mathfrak{f}(G) = \pm 1 \Leftrightarrow G$ solvable of odd order; on the other hand, one would like to conjecture that $\mathfrak{f}(G)$ is always generated by -1 as an $\Omega(G)$-module, but unfortunately this is not true already for the nonabelian group of order 10. But still it seems interesting and promising to study the structure of $\mathfrak{f}(G)$ as an $\Omega(G)$-module. Let me add one more elementary result in this direction:

PROPOSITION 3.5. *If G is of odd order, then $\mathfrak{f}(G)$ is a semisimple $\Omega(G)$-module, which is generated as a $\Omega(G)$-module by one element. In general $\mathfrak{f}(G)$ is isomorphic to a submodule of $\mathrm{Hom}_z(\Omega(G), F_2)$, which is an $\Omega(G)$-module by:*

$$f \in \mathrm{Hom}_z(\Omega(G), F_2), \quad x, y \in \Omega(G) \Rightarrow (xf)(y) = f(xy).$$

4. **Adjoints to restriction.** Let $U \leq G$ be a subgroup of G. Restriction to U defines a functor from the category of G-sets to the category of U-sets, which maps sums into sums and products into products, more generally: direct and inverse limits into direct resp. inverse limits. Thus it defines ring homomorphism $\Omega(G) \to \Omega(U)$, especially a multiplicative map $\mathfrak{f}(G) \to \mathfrak{f}(U)$ which is easily seen to be an $\Omega(G)$-module-homomorphism, if $\mathfrak{f}(U)$ is considered as an $\Omega(G)$-module via $\Omega(G) \to \Omega(U)$.

The restriction functor has a left and a right adjoint. The left adjoint maps a U-set S onto the G-set $G \times_U S$, the set of U-orbits in $G \times S$ for the U-action: $u \in U$, $g \in G$, $s \in S \Rightarrow u(g,s) = (gu^{-1}, us)$ with the well-defined G-action $g'(g,s)^- = (g'g,s)^-$ ((g,s)$^-$ the U-orbit of $(g,s) \in G \times S$).

This map is easily seen to be additive, but generally not multiplicative. Instead, by the adjointness it fulfills some kind of Frobenius reciprocity law, i.e. the additive map $\Omega(U) \to \Omega(G)$, defined by $[S] \mapsto [G \times_U S]$, is an $\Omega(G)$-module-homomorphism if one considers $\Omega(U)$ as an $\Omega(G)$-module via the restriction map $\Omega(G) \to \Omega(U)$. Thus $\Omega(\cdot)$ becomes a Frobenius functor in the sense of Lam (see [4]), and one can prove also that it is a G-functor in the sense of Green (see notes to his lecture in this volume). Moreover, one can prove that in the category of G-functors for a fixed group G the G-functor: $U \mapsto \Omega(U)$ ($U \leq G$) is the initial object which explains the role of $\Omega(G)$ in representation theory, as observed by Conlon [1] (see also [3]). The right adjoint to restriction is given by $S \mapsto {}^G S$, where ${}^G S$ is the G-set of set-theoretic sections in the canonical epimorphism $c: G \times_U S \to G/U: (g,s) \mapsto gU$ (G/U the G-set of left cosets gU of U in G), i.e. ${}^G S = \{f: G/U \to G \times_U S | cf = \mathrm{Id}_{G/U}\}$ with the G-action: $g \in G$, $f \in {}^G S$, $x \in G/U \Rightarrow (gf)(x) = g(f(g^{-1}x))$.

This map is not additive but multiplicative and one can prove

PROPOSITION 4.1. *The map* $\Omega^+(U) \to \Omega(G): [S] \mapsto [{}^G S]$ *is algebraic of degree* $(G:U)$; *thus it extends to a multiplicative map*: $\Omega(U) \to \Omega(G)$, *which is an* $\Omega^+(G)$-*semimodule-homomorphism, if one considers* $(\Omega(U), \cdot)$ *as an* $\Omega^+(G)$-*semimodule via the restriction map* $\Omega^+(G) \to \Omega^+(U)$.

Especially one gets maps $\mathfrak{f}(U) \to \mathfrak{f}(G)$, which might be useful in the study of the structure of $\mathfrak{f}(G)$. But there are other interesting applications of this construction, as we shall see in the next two sections.

5. **On the transfer homomorphism.** Let R be a commutative ring with $1 \in R$. An RG-module M is a finitely presented left R-module, on which G acts from the left R-linearly. RG-modules form a category **RG-mod** and one has a natural functor from the category **G-set** of G-sets into this category which maps any G-set S onto the RG-module $R[S]$, which is the free R-module generated by S, with the G-action extended R-linearly from S to $R[S]$. For a subgroup $U \leq G$ we have again a restriction functor

RG-mod → RU-mod

which makes the diagram

$$\begin{array}{ccc} RG\text{-mod} & \to & RU\text{-mod} \\ \uparrow & & \uparrow \\ G\text{-set} & \to & U\text{-set} \end{array}$$

commutative and has an adjoint RU-mod → RG-mod: $M \to RG \otimes_{RU} M$ which is a left and a right adjoint at the same time.

Moreover, the diagram

$$\begin{array}{ccc} RG\text{-mod} & \leftarrow & RU\text{-mod} \\ \uparrow & & \uparrow \\ G\text{-set} & \leftarrow & U\text{-set} \end{array}$$

is commutative if the lower arrow is the left adjoint to restriction on **G-set**, but it is not commutative if the lower arrow is the right adjoint to the restriction. Thus one might ask for a construction of RG-modules out of RU-modules, which makes this diagram commutative if the lower arrow is the right adjoint. There is indeed such a construction: Let M be a RU-module and $g_1, \ldots, g_n$ be a complete system of coset-representations of G/U. Then $RG \otimes_{RU} M = \bigoplus (g_i \otimes M)$ and G acts on $RG \otimes_{RU} M$ by permuting the blocks $g_i \otimes M = M_i$. Thus instead of taking the sum $\bigoplus M_i$, one can consider the tensor product $\bigotimes_R M_i = {}^G M$ and define an action of G on ${}^G M$ by the way an element $g \in G$ maps $M_i = g_i \otimes M$ into $M_j = g_j \otimes M = g g_i \otimes M$.

This way ${}^G M$ becomes an RG-module and one can extend this construction in an obvious way to a functor RU-mod → RG-mod which makes the diagram

$$\begin{array}{ccc} RU\text{-mod} & \to & RG\text{-mod} \\ \uparrow & & \uparrow \\ U\text{-set} & \to & G\text{-set} \end{array}$$

commutative, where the lower arrow now is the right adjoint to the restriction. Moreover, we have

$$^G(M \otimes_R N) \cong {}^G M \otimes_R {}^G N;$$

thus this functor is multiplicative, but in general not additive.

The isomorphism classes of RG-modules form a halfring $a^+(G;R)$ with $[M] + [N] = [M \oplus N]$, $[M] \cdot [N] = [M \otimes_R N]$. Let $a(G;R)$ be the associated universal ring. The functor **G-set** → **RG-mod** defines a ring homomorphism $\Omega(G) \to a(G;R)$, restriction gives a ring homomorphism $a(G;R) \to a(U;R)$, and the usual induction makes $a(\cdot;R)$ to a G-functor or Frobenius functor.

Moreover $M \mapsto {}^G M$ defines a multiplicative map $a^+(U;R) \to a(G;R)$, which again is algebraic of degree $(G:U)$; thus we get a unique algebraic multiplicative map $a(U;R) \to a(G;R)$.

Now let $R = \mathbf{C}$ and consider the subset $a_1(G,\mathbf{C})$ in $a(G,\mathbf{C})$ which consists of elements represented by $\mathbf{C}G$-modules of $\mathbf{C}$-dimension 1. $a_1(G,\mathbf{C})$ is a multiplicative subgroup of $a(G,C)$ isomorphic to $\hat{G} = \text{Hom}(G,C^{\times})$, the dual group of G, resp. $G/[G,G]$.

Our map defines a multiplicative homomorphism $a_1(U;C) \cong \hat{U} \to a_1(G,C) \cong \hat{G}$, thus dualizing a map $G/[G,G] \to U/[U,U]$, which turns out to be the transfer homomorphism. Thus our construction gives a representation theoretic generalization of the transfer homomorphism and at the same time perhaps a new and more invariant definition of transfer. We will give an application of this construction, concerning induction theorems for relative Grothendieck rings.

6. **Relative Grothendieck rings.** Let $\mathfrak{A}$ be a nonvoid family of subgroups of G. A sequence $E: O \to M' \to M \to M'' \to O$ of RG-modules is called $\mathfrak{A}$-split if for any $U \in \mathfrak{A}$ the sequence, restricted to U, splits. Let $a(G,\mathfrak{A},R)$ be the quotient ring of $a(G;R)$ modulo the ideal $i(G,\mathfrak{A},R)$, generated by the Euler characteristics $\chi_E = [M'] - [M] + [M''] \in a(G;R)$ of all $\mathfrak{A}$-split sequences E of RG-modules. To make $a(G,\mathfrak{A};R)$ to a Frobenius functor we define $\bar{\mathfrak{A}} = \{U' \leq G | \exists g \in G, U \in \mathfrak{A}$ with $gU'g^{-1} \subseteq U\}$. For $H \leq G$ let $H \cap \bar{\mathfrak{A}}$ be the family $\{H \cap U | U \in \bar{\mathfrak{A}}\}$ of subgroups of H. Then the restriction $a(G;R) \to a(H;R)$ and the additive induction $a(H;R) \to a(G;R)$ map $i(G;\mathfrak{A};R)$ into $i(H, H \cap \bar{\mathfrak{A}};R)$, resp. $i(H, H \cap \bar{\mathfrak{A}};R)$ into $i(G,\mathfrak{A};R)$; thus these maps induce maps $a(G,\mathfrak{A};R) \to a(H, H \cap \mathfrak{A};R)$, $a(H, H \cap \mathfrak{A};R) \to a(G,\mathfrak{A};R)$, which make $a_{\mathfrak{A}R}: H \mapsto a(H, H \cap \bar{\mathfrak{A}};R)$ a Frobenius- and even a G-functor.

The following theorems state some results concerning the defect-basis (see Green, this volume) of $a_{\mathfrak{A},R}$, resp. $P \otimes_Z a_{\mathfrak{A},R}$ for some commutative ring P. Let $\mathscr{D}_p(\mathfrak{A},R)$ be the defect-basis of $P \otimes_Z a_{\mathfrak{A},R}$.

THEOREM 6.1. (a) $\mathscr{D}_p(\mathfrak{A},R) = \mathscr{D}_p(\mathfrak{A}_R,R)$ with $\mathfrak{A}_R = \{U \in \bar{\mathfrak{A}} \| U |$ a p-power with $pR \neq R\}$ resp. $\mathfrak{A}_R = \{E\}$ if $pR = R$ for all primes p. (Maschke, Gaschütz, D. G. Higman.)

(b) If $\mathfrak{B}$ is another family of subgroups of G, then $\mathscr{D}_p(\mathfrak{A} \cup \mathfrak{B},R) = \mathscr{D}_p(\mathfrak{A},R) \cup \mathscr{D}_p(\mathfrak{B},R)$.

(c) $\mathscr{D}_p(\mathfrak{A},R) = \cup_{\mathfrak{m}} \mathscr{D}_p(\mathfrak{A},R_{\mathfrak{m}})$ where $\mathfrak{m}$ runs through all maximal ideals of R.

THEOREM 6.2. If R is a local ring with residue class characteristic p, then

$$\mathfrak{C}_p \mathfrak{A} \subseteq \mathscr{D}_Q(\mathfrak{A},R) \subseteq \mathfrak{C}_p \mathfrak{A} \cup \mathfrak{C},$$

where $\mathfrak{C}$ is the family of cyclic subgroups of G and

$$\mathfrak{C}_p \mathfrak{A} = \{V \leq G | V_p \trianglelefteq V, V_p \in \mathfrak{A}, V/V_p \text{ cyclic}\},$$

V_p the p-Sylow subgroup of V. We have equality on the left, if R is a field, and equality on the right, if R can be mapped into a field of characteristic 0.

THEOREM 6.3. If R,p is as in Theorem 6.2 and $1/p \in P$, then $\mathscr{D}_p(\mathfrak{A},R) \subseteq \{V \leq G | \exists N \trianglelefteq V, N \in \mathfrak{C}_p \mathfrak{A} \cup \mathfrak{C}$ and V/N a q-group for some q with $qP \neq P\}$. Equality holds if $R = \mathbf{Z}_P \subseteq \mathbf{Q}$ or $R = \hat{\mathbf{Z}}_p$. In case R is a field, it is enough to require $N \in \mathfrak{C}_p \mathfrak{A}$.

THEOREM 6.4. *For any R and $\mathfrak{A} = \{G\}$ we have $\mathscr{D}_z(\{G\}, R) \subseteq \{V \leq G | \exists N \trianglelefteq V, N$ a p-group for some p with $pR \neq R$ and V/N hyperelementary$\}$. If $R \subseteq \mathbf{Q}$, equality holds. If R is a $\mathbf{Z}[\zeta]$-algebra, where ζ is a primitive nth root of unity and $n = \exp G$, then equality holds if one requires V/N to be elementary.*

There are also similar generalizations of the Berman-Witt Theorem.

These theorems contain many of the results of Artin, Brauer, Conlon and Swan on induction. The proofs are rather complicated. One important step is to prove: If $H \trianglelefteq G$ and if all groups in $\mathfrak{A} = \bar{\mathfrak{A}}$ are contained in H, then the multiplicative induction map $a(H; R) \to a(G; R)$, considered in §5, induces a well-defined map

$$a(H, \mathfrak{A}; R) \to a(G, \mathfrak{A}; R).$$

It is an open question whether it induces always (i.e. with $H \leq G$ arbitrary and $\mathfrak{A}$ arbitrary) a well-defined map: $a(H, H \cap \bar{\mathfrak{A}}; R) \to a(G, \mathfrak{A}; R)$.

LITERATURE

1. S. B. Conlon, *Decompositions induced from the Burnside algebra*, J. Algebra **10** (1968), 102–122. MR **38** #5945.

2. A. Dress, *A characterization of solvable groups*, Math. Z. **110** (1969), 213–217. MR **40** #1491.

3. A. Dress and M. Küchler, *Zur Darstellungstheorie endlicher Gruppen*, Vorlesungsausarbeitung, Bielefeld, Fakultät für Mathematik, Universität.

4. T.-Y. Lam, *Induction theorems for Grothendieck groups and Whitehead groups of finite groups*, Ann. Sci. École Norm. Sup. (4) **1** (1968), 91–148. MR **38** #217.

5. J. Riordan, *An introduction to combinatorial analysis*, Wiley, New York; Chapman & Hall, London, 1958, pp. 34–38. MR **20** #3077.

6. L. Solomon, *The Burnside algebra of a finite group*, J. Combinatorial Theory **2** (1967), 603–615. MR **35** #5528.

UNIVERSITÄT BIELEFELD, GERMANY

Some Decomposable Sylow 2-Subgroups and a Nonsimplicity Condition

Paul Fong

Let G be a finite group with Sylow 2-subgroup P. Assume the elements of P are fused as in a direct product; what are sufficient conditions for the nonsimplicity of G? We have the following:

LEMMA. *Let G be a finite group with Sylow 2-subgroup P. Suppose there exist involutions j_1, j_2 in P, and a nonprincipal irreducible character χ of G satisfying the following conditions:*
  (1) *If j'_1, j'_2 are involutions in P fused to j_1, j_2 respectively, then $j'_1 j'_2$ is fused to $j_1 j_2$.*
  (2) $\chi(j_1 j_2 v) = \chi(j_1 j_2) \neq 0$ *for all 2-regular elements v in $C(j_1 j_2)$.*
  (3) $|\chi(j_1 j_2)| \geq |\chi(j_1)|$ *or* $|\chi(j_2)|$.
Then G is not a nonsolvable simple group.

PROOF. An essential part of the proof is due to G. Higman and G. Glauberman. Let s_1, s_2 be any conjugates of j_1, j_2 respectively, and let p be the 2-part of $s_1 s_2$. Since $\langle s_1, s_2 \rangle$ is dihedral, there is a conjugate s'_2 of s_2 in $\langle s_1, s_2 \rangle$ such that $s_1 s'_2 = p$. By Sylow's Theorem we can choose g in G such that $\langle s_1, s'_2 \rangle^g \leq P$. (1) implies that $(s_1 s'_2)^g$ and $j_1 j_2$ are conjugate, or equivalently, p and $j_1 j_2$ are conjugate. Let K_1^1, K_2^1 be the conjugate classes of G containing j_1, j_2 respectively, and let $K_1, \ldots, K_t$ be the classes of G in the section of $j_1 j_2$. We have just shown that

$$[K_1^1][K_2^1] = \sum_{j=1}^{t} a_j [K_j],$$

where the a_j are nonnegative integers, and $[K]$ for any subset K of G is the sum

---

AMS 1969 *subject classifications*. Primary 2025.

$\sum g$, $g \in K$, in the group ring $Z[G]$. If ζ is any irreducible character of G, and g_j is a representative of K_j, then

$$|K_1^1||K_2^1|\zeta(j_1)\zeta(j_2) = \sum_{j=1}^{t} a_j |K_j|\zeta(g_j)\zeta(1).$$

Setting $\zeta = 1$ and $\zeta = \chi$ respectively and using (2), we find that $\chi(1)\chi(j_1 j_2) = \chi(j_1)\chi(j_2)$. Since $\chi(j_1 j_2) \neq 0$, it follows from (3) that $|\chi(j_i)| \geq \chi(1)$ for $i = 1$ or 2. Thus G is not simple.

As an application of the preceding, we have the following:

COROLLARY. *If P is the direct product of two dihedral groups of unequal order, then P cannot be the Sylow 2-subgroup of a simple group.*

The conditions of the lemma are 2-local in the sense that they involve fusion of elements of P and values of irreducible characters of G on 2-singular sections. The required character can then be constructed by a combination of the exceptional character theory and block theory. It is quite likely that the general case of two dihedral groups, of equal or unequal order, can be settled by the methods of signalizer functors introduced by D. Gorenstein.

UNIVERSITY OF ILLINOIS AT CHICAGO CIRCLE

Characters and Orthogonality in Frobenius Algebras

T. V. Fossum

Summary. In this paper we develop a matrix-free, module-theoretical proof of orthogonality relations for the irreducible characters of a Frobenius algebra.

A *Frobenius algebra* over a field K is a pair (A,φ) such that A is a finite-dimensional K-algebra, and $\varphi: A \to A^* = \mathrm{Hom}_K(A,K)$ is a left A-module isomorphism. Here A^* is a left A-module by defining $(af)(b) = f(ba)$ for all $a, b \in A, f \in A^*$. If $(a_i), (b_i)$ is an ordered pair of bases for a Frobenius algebra (A,φ) such that $\varphi(b_i)(a_j) = \delta_{ij}$ for all i,j, we say the bases are φ-*dual*.

Let A be a finite-dimensional K-algebra, K a field, and let M be a simple left A-module. Set $A_M = \{a \in A : aM = 0\}$. Then A_M is a maximal 2-sided ideal of A, and if N is a simple left A-module such that $A_M N = 0$, then $N \cong M$.

LEMMA. *Let (A,φ) be a Frobenius algebra, and let M be a simple left A-module with character χ. Set $\chi^* = \varphi^{-1}(\chi)$. Then $A_M \chi^* = 0$. If N is a simple left A-module such that $\chi^* N \neq 0$, then $N \cong M$.*

THEOREM (ORTHOGONALITY RELATIONS). *Let (A,φ) be a Frobenius algebra with φ-dual bases $(a_i), (b_i)$. Let M,N be simple left A-modules with characters χ, ζ, respectively. If $\sum_i \chi(a_i)\zeta(b_i) \neq 0$, then $M \cong N$.*

LEMMA. *Let (A,φ) be a Frobenius algebra with φ-dual bases $(a_i), (b_i)$, and let ρ be the character of the left regular module $_A A$. Set $\rho^* = \varphi^{-1}(\rho)$. Then $\rho^* = \sum_i a_i b_i$.*

THEOREM. *Let (A,φ) be a Frobenius algebra with φ-dual bases $(a_i), (b_i)$, and let ρ be the character of $_A A$. If χ is the character of a simple left A-module M, then*

AMS 1969 *subject classifications.* Primary 1640; Secondary 2080.

$\chi(\sum_i a_i b_i) = \chi(\rho^*) = m\sum_i \chi(a_i)\chi(b_i)$, where m is the multiplicity of M as a composition factor of $_AA$.

In the case of a group-algebra KG of a finite group G over a splitting field K of characteristic zero, we have that $(g^{-1}), (g)$ are φ-dual bases for (KG,φ), where $\varphi: KG \to KG^*$ is the usual isomorphism, and therefore $\sum_{g\in G} g^{-1}g = |G|\cdot 1$. Moreover, KG is semisimple, and any simple module M with character χ appears exactly $\chi(1)$ times as a composition factor of KG. We then conclude the following.

COROLLARY. *Let G be a finite group, K a splitting field of characteristic zero. Let χ be the character of a simple left KG-module. Then $|G| = \sum_{g\in G} \chi(g^{-1})\chi(g)$.*

UNIVERSITY OF UTAH

The Number of Conjugacy Classes in a Finite Group

P. X. Gallagher

Let $k(G)$ denote the number of conjugacy classes (= number of irreducible characters) of a finite group G. Ernest [1] has shown that for each subgroup H,

(1) $$k(G) \leqq (G:H)k(H).$$

If there is equality in (1), then H is a normal subgroup. Sah [3] has shown that for each normal subgroup N, the equation

(2) $$k(G) = |G/N| \cdot k(N)$$

is equivalent to each of the following:
  (a) $G = C(\sigma)N$, for each $\sigma \in G$;
  (b) G/N is abelian, and each character of N extends to a character of G.

THEOREM 1. *If N is a normal subgroup of G, then*

(3) $$k(G) \leqq k(G/N)k(N),$$

and equality is equivalent to each of the following:
  (a) *$C(\sigma \bmod N) = C(\sigma)N$, for each $\sigma \in G$;*
  (b) *each character of N extends to each subgroup $N\langle\sigma,\tau\rangle$ with $[\sigma,\tau] \in N$.*
If the Sylow p-subgroups of G/N are abelian, for each p dividing $|N|$, then equality is equivalent to:
  (c) *each character of N extends to a character of G.*

Our first proof of (3) involves classes and leads to the condition (a) for equality. The second involves characters and leads to (b). Condition (c) is derived from (b)

AMS 1970 *subject classifications.* Primary 20C15.

by induction in the case G/N abelian, and by a cohomology argument in the general case.

The argument leading to (b) is based on Clifford's theorem and a variant of a result of Schur [4] and Ernest [2]: Let ψ be an irreducible character of N. Assume ψ is invariant under G. If G/N is cyclic, then ψ extends to a character of G, so, in general, ψ extends to a character of each subgroup $N\langle\sigma\rangle$. If the extension to $N\langle\sigma\rangle$ is invariant under C (σ mod N), then σ is good for ψ. Goodness is independent of the choice of the extension, and depends only on ψ and the conjugacy class of σN in G/N.

THEOREM 2. *The number of distinct irreducible components of the induced character ψ^G is equal to the number of classes of G/N which are good for ψ.*

In response to a problem posed at the end of the talk, E. C. Dade gave an example in which equality holds in (3) but not every character of N extends to a character of G.

REFERENCES

1. J. A. Ernest, *Central intertwining numbers for representations of finite groups*, Trans. Amer. Math. Soc. **99** (1961), 499–508. MR **23** #A2467.

2. ———, *Embedding numbers for finite groups*, Proc. Amer. Math. Soc. **13** (1962), 567–570. MR **25** #1218.

3. C.-H. Sah, *Automorphisms of finite groups*, J. Algebra **10** (1968), 47–68. MR **37** #5287.

4. I. Schur, *Über die Darstellungen endlichen Gruppen durch gebrochene lineare Substitutionen*, J. Reine Angew. Math. **127** (1904), 20–50.

BARNARD COLLEGE, COLUMBIA UNIVERSITY

Sylow 2-Subgroups with non-Elementary Centers

David M. Goldschmidt

I would like to discuss briefly some results obtained over the past several years on Sylow 2-subgroups of simple groups. Proofs will appear elsewhere.

Let $\text{Syl}_p(G)$ denote the set of Sylow p-subgroups of a finite group G. For a p-group P and a nonnegative integer n, let

$$\mho^n(P) = \langle x^{p^n} | x \in P \rangle.$$

It is well known that Sylow 2-subgroups of simple groups have very restricted structure as abstract 2-groups. For example, as an instant corollary of John Walter's recent classification of simple groups with abelian Sylow 2-groups [3], we see that abelian Sylow 2-subgroups of simple groups are elementary abelian. We can generalize this result as follows:

THEOREM 1. *Let G be a finite simple group, with $T \in \text{Syl}_2(G)$. If T has nilpotence class $n > 1$, then $Z(T)$ has exponent at most 2^{n-1}.*

This bound is sharp as is seen, for example, in the groups $\text{PSL}(3,q)$ where $q = 2^{n-1}a, n > 2$, and a is odd. Here, $T \cong Z_{2^{n-1}} \wr Z_2$, a 2-group of nilpotence class n whose center has exponent 2^{n-1}.

Theorem 1 is a corollary of the following two results:

THEOREM 2. *Let G be a finite group with $T \in \text{Syl}_2(G)$. If T has nilpotence class n, then $\mho^{n-1}(Z(T))$ is weakly closed in T with respect to G.*

AMS 1970 *subject classifications.* Primary 20D05, 20G40.

THEOREM 3. *Let G be a finite group with $T \in \mathrm{Syl}_2(G)$. Suppose $W \subseteq \mho^1(Z(T))$ and W is weakly closed in T with respect to G. Then $W \subseteq O_{2',2}(G)$.*

For odd primes, there is an analogue of Theorem 2, but no analogue of Theorem 3.

Theorem 2 can be sharpened by introducing the Thompson subgroup, defined as follows:

For a p-group P, let

$$p^{d(P)} = \max\{|A| \,\big|\, A \subseteq P, A \text{ abelian}\},$$
$$\mathscr{A}(P) = \{A \subseteq P | A \text{ abelian}, |A| = p^{d(P)}\},$$
$$J(P) = \langle A | A \in \mathscr{A}(P) \rangle.$$

THEOREM 4. *Let G be a finite group with $P \in \mathrm{Syl}_p(G)$. If $\mho^1(Z(P)) \subseteq O_p(G)$, then*

$$G = N_G(J(P)) \cdot C_G(\mho^1(Z(P))).$$

For subsets $A, B \subseteq G$, write $A \sim_G B$ whenever $A^g = B$ for some $g \in G$. Theorem 4 yields the following fusion result:

THEOREM 5. *Let G be a finite group with $P \in \mathrm{Syl}_p(G)$. Let $N = N(J(P))$, $C = C_G(\mho^1(Z(P)))$. Let $\approx$ be the equivalence relation on subsets of P generated by the relations $\sim_N$ and $\sim_C$. Then for $A, B \subseteq P$; $A \approx B$ iff $A \sim_G B$.*

This result sharpens Theorem 3:

COROLLARY 6. *Let G be a finite group with $T \in \mathrm{Syl}_2(G)$. Suppose $W \subseteq \mho^1(Z(T))$ and $W \trianglelefteq N_G(J(T))$. Then $W \subseteq O_{2',2}(G)$.*

Further progress can be made by improving Theorem 4 with an analogue of the Thompson "three against two" theorem [2]. To do this, we introduce another characteristic subgroup of a p-group P as follows:

Let $\mathscr{A}_1(P) = \{A \subseteq P | A \text{ abelian}, |A| \geq p^{d(P)-1}\}$,
$J_1(P) = \langle A | A \in \mathscr{A}_1(P) \rangle$.

THEOREM 7. *Let G be a finite group with $P \in \mathrm{Syl}_p(G)$. Suppose $\mho^2(Z(P)) \subseteq O_p(G)$; then*

$$G = N_G(J_1(P)) \cdot C_G(\mho^2(Z(P))).$$

Moreover, if G is solvable and $O_{p'}(G) = 1$, then

$$G = N_G(J(P)) \cdot N_G(\mho^2(Z(J_1(P)))).$$

In view of previous results of Glauberman [1] the second factorization is uninteresting unless $p = 2$. I originally stated a stronger version of Theorem 7, for which I had an incorrect proof. I would like to thank Professor Glauberman for pointing out this error.

Based on Theorem 7 and further results which are too technical to state here, one is led to the following:

CONJECTURE. Let G be a finite simple group with $T \in \mathrm{Syl}_2(G)$. Then $C_G(\mho^2(Z(T)))$ is nonsolvable.

In summary, it seems clear that for simple groups G with $T \in \mathrm{Syl}_2(G)$, the case $\mho^1(Z(T)) > 1$ is anomalous.

This case occurs in certain Chevalley groups of odd characteristic. From the point of view of Chevalley groups, however, the hypothesis is somewhat artificial. One would prefer a hypothesis which somehow "captures" Sylow 2-subgroups of all Chevalley groups of odd characteristic.

BIBLIOGRAPHY

1. G. Glauberman, *A characteristic subgroup of a p-stable group*, Canad. J. Math. **20** (1968), 1101–1135. MR **37** #6365.

2. J. G. Thompson, *Factorizations of p-solvable groups*, Pacific J. Math. **16** (1966), 371–372. MR **32** #5735.

3. J. Walter, *The characterization of finite groups with abelian Sylow 2-subgroups*, Ann. of Math. (2) **89** (1968), 405–514.

YALE UNIVERSITY

Axiomatic Representation Theory

J. A. Green

1. **G-functors.** This lecture is intended to describe a new branch of the representation theory of finite groups. Proofs of theorems and other details will be found in [5].

Given a finite group G and a commutative ring k with identity element, we shall define *G-functors over* k. These take the place of *representations of G over k* in classical representation theory. In particular, G-functors "represent" the subgroup structure of G, as we shall now explain.

First we need some terminology. A *k-algebra* is to be a *k*-module P, equipped with a bilinear multiplication $(\alpha,\beta) \mapsto \alpha.\beta$. A *k-map* $\theta: P \to P'$, where P,P' are k-algebras, is to be a k-module homomorphism from P into P'; thus k-maps need not be connected in any way with the multiplications in P,P'. We shall say that $\theta: P \to P'$ is *multiplicative* if $(\alpha.\beta)\theta = \alpha\theta.\beta\theta$ for all α,β in P. The identity map on P is denoted $\mathrm{id}(P)$.

DEFINITION. A G-functor over k is defined to be a quadruple

$$A = (A, T, R, C),$$

where A, T, R, C are families of the following kinds:

$A = (A_H)$ gives, for each subgroup H of G (notation $H \leq G$), a k-algebra A_H.

$T = (T_{H,K})$ and $R = (R_{K,H})$ give, for each pair (H,K) of subgroups of G such that $H \leq K$, the respective k-maps

$$T_{H,K}: A_H \to A_K \quad \text{and} \quad R_{K,H}: A_K \to A_H.$$

$C = (C_{H,g})$ gives, for each pair (H,g) where H is a subgroup, and g an element of G, the k-map

---

AMS 1970 *subject classifications.* Primary 20C99, 20J99, 20–02.

$$C_{H,g}: A_H \to A_{H^g} \qquad (H^g = g^{-1}Hg).$$

These families of k-algebras and k-maps must satisfy the following

AXIOMS FOR G-FUNCTORS. (In these axioms, D,H,K,L are any subgroups of G; g,g' are any elements of G; α,α' are any elements of A_H; β,β' are any elements of A_K.)

(a) $T_{H,H} = \mathrm{id}(A_H)$, $T_{H,K}T_{K,L} = T_{H,L}$ if $H \leq K \leq L$.
(b) $R_{H,H} = \mathrm{id}(A_H)$, $R_{K,H}R_{H,D} = R_{K,D}$ if $K \geq H \geq D$.
(c) $C_{H,g}C_{H^g,g'} = C_{H,gg'}$.
(d) $C_{H,h} = \mathrm{id}(A_H)$ if $h \in H$.
(e) $T_{H,K}C_{K,g} = C_{H,g}T_{H^g,K^g}$.
(f) $R_{K,H}C_{H,g} = C_{K,g}R_{K^g,H^g}$.
(g) *Mackey Axiom*. If $H \leq L$, $K \leq L$ and if Γ is a transversal of the (H,K) double cosets in L, then

$$T_{H,L}R_{L,K} = \sum_{g \in \Gamma} C_{H,g}R_{H^g, H^g \cap K}T_{H^g \cap K, K}.$$

(h) *Frobenius Axiom*. If $H \leq K$ then

$$\alpha T_{H,K} \cdot \beta = (\alpha \cdot \beta R_{K,H})T_{H,K} \text{ and } \beta \cdot \alpha T_{H,K} = (\beta R_{K,H} \cdot \alpha)T_{H,K}.$$

(i) $C_{H,g}$ is multiplicative.

(M) $R_{K,H}$ is multiplicative.

This rather formidable list of axioms can be shortened a bit by introducing a suitable *subgroup category* of G [5, §1]. However, Axioms (a)–(M) provide in practice the most convenient form for the definition of G-functors. It is curious that axiom (M), which appears very naturally in the examples to be given below, is not necessary in the proofs of our two main general theorems on G-functors. We should mention here T.-Y. Lam's *Frobenius functors*, defined in [6], which have many features in common with G-functors.

2. **Examples of G-functors.** In these examples, H and K are arbitrary subgroups of G such that $H \leq K$; g is an arbitrary element of G. We shall describe in each case the algebras A_H and the maps $T_{H,K}$, $R_{K,H}$ and $C_{H,g}$ which describe the G-functor $A = (A,T,R,C)$, but we do not give the verification of Axioms (a)–(M) which is usually rather lengthy.

EXAMPLE 1. *The character ring functor on G*. We take $k = Z$, and define a G-functor as follows.

A_H is the *character ring* of H, i.e. $A_H = \{\sum z_i \psi_i | z_i \in Z\}$, where the ψ_i are the irreducible ordinary characters of H. This is a Z-algebra, under the usual multiplication of characters.

$T_{H,K}$ takes $\psi \in A_H$ to the *induced character* $\psi^K \in A_K$.

$R_{K,H}$ takes $\chi \in A_K$ to its *restriction* $\chi_H \in A_H$.

$C_{H,g}$ takes $\psi \in A_H$ to the *conjugate character* $\psi^g \in A_{H^g}$, where $\psi^g(u) = \psi(gug^{-1})$, for any $u \in H^g = g^{-1}Hg$.

EXAMPLE 2. *The cohomology ring functor on G.* Let k be any commutative ring with 1; we regard k as trivial G-module. Then the following define a G-functor over k.

A_H is the *Tate cohomology ring* $\hat{H}^*(H,k)$.

$T_{H,K}: A_H \to A_K$ is the *Eckmann transfer*.

$R_{K,H}$, $C_{H,g}$ are most easily defined as the maps induced, in the relevant Tate cohomology rings, by the group homomorphisms inc: $H \to K$, conj: $H^g \to H$ (conj is the map $u \mapsto gug^{-1}$). We are using here the fact that $H \to \hat{H}^*(H,k)$ extends to a contravariant functor on the category of groups and group homomorphisms. (The maps $R_{K,H}$, $C_{H,g}$ in Examples 1 and 3 can be constructed in an analogous way.)

EXAMPLE 3. *The relative Grothendieck ring functors on G.* Take a commutative ring R with 1, and a fixed normal subgroup Y of G. Then we can make a G-functor over Z, which can be regarded as a generalisation of the character ring functor on G, Example 1.

A_H is the *relative Grothendieck ring* $a(H, H \cap Y)$ of Lam-Reiner, defined on the category of finitely-generated RH-modules, relative to RH-exact sequences which are $R(H \cap Y)$-split (see [7]).

$T_{H,K}$ takes $[L] \mapsto [L^K]$ (L an RH-module).

$R_{K,H}$ takes $[M] \mapsto [M_H]$ (M an RK-module).

$C_{H,g}$ takes $[L] \mapsto [L^g]$ (L an RH-module).

It is necessary, of course, to check that these definitions for $T_{H,K}$, $R_{K,H}$ and $C_{H,g}$ are consistent with the relations which define the relative Grothendieck rings in question. It is for this reason that we take Y to be a *normal* subgroup of G. In case $Y = \{1\}$, A_H is the ordinary Grothendieck ring; in case $Y = G$, A_H is the representation ring of H.

EXAMPLE 4. *The G-functor defined by a G-algebra.* For any G and k, a G-*algebra over k* is a k-algebra A, on which G acts ($g \in G$ acts on $\alpha \in A$ to give α^g), so that A becomes a kG-module, and also $(\alpha.\beta)^g = \alpha^g.\beta^g$, for all $\alpha, \beta \in A$, $g \in G$. Then we make the G-functor over k defined by A as follows:

$A_H = \{\alpha \in A | \alpha^h = \alpha \text{ for all } h \in H\}$.

$T_{H,K}$ takes $\alpha \in A_H$ to $\sum \alpha^v$, sum over an H-transversal $\{v\}$ of K. If $H \leq K$, then clearly $A_K \leq A_H$, so we take

$R_{K,H} = \text{inc}: A_K \to A_H$.

$C_{H,g}$ takes $\alpha \in A_H$ to α^g, which evidently lies in A_{H^g}.

It is this last type of G-functor which seems most closely involved with modular representation theory; we shall give two instances of this later on.

EXAMPLE 5. G-functors over k (G,k given) are the objects of a category $A_k(G)$, whose morphisms are "natural transformations" between G-functors [5, §1.5]. We can define subobjects and quotient objects in this category. For example, a G-functor A' is a subfunctor of a G-functor A, if for each $H \leq G$ the algebra A'_H is a subalgebra of A_H, and if this family (A'_H) is closed to the "operations" $T_{H,K}, R_{K,H}$ and $C_{H,g}$ (e.g. $A'_H T_{H,K} \leq A'_K$, whenever $H \leq K \leq G$), and if $T'_{H,K}, R'_{K,H}$ and $C'_{H,g}$ are the appropriate restrictions of $T_{H,K}, R_{K,H}$ and $C_{H,g}$.

3. The defect base of a G-functor.

Let $A = (A, T, R, C)$ be any G-functor over k. If H is any subgroup of G, write $s(H)$ for the set of all non-$\emptyset$ sets of subgroups of H. Remember that $U \leq H$ means U is a subgroup of H.

DEFINITION. If $U \leq H$, define $A_{U,H} = A_U T_{U,H}$. If $\mathfrak{U} \in s(H)$, define $A_{\mathfrak{U},H} = \sum_{U \in \mathfrak{U}} A_{U,H}$.

By the Frobenius axiom (h), $A_{U,H}$ and $A_{\mathfrak{U},H}$ are ideals of A_H.

DEFINITION. If $\mathfrak{U} \in s(H)$, define

$$j_H \mathfrak{U} = \{S \leq H | S^h \leq U \text{ for some } h \in H, U \in \mathfrak{U}\}.$$

We call $j_H \mathfrak{U}$ the j_H-closure of $\mathfrak{U}$. If $j_H \mathfrak{U} = \mathfrak{U}$, we say $\mathfrak{U}$ is j_H-closed. It is a trivial calculation that, for any $\mathfrak{U} \in s(H)$,

$$A_{\mathfrak{U},H} = A_{j_H \mathfrak{U}, H}.$$

So we have, in the k-algebra A_H, a family or "filtration" $(A_{\mathfrak{U},H})$ of ideals; the filtration index $\mathfrak{U}$ runs over $s(H)$, or, if we want to economise, the last formula shows that we may restrict $\mathfrak{U}$ to j_H-closed elements of $s(H)$.

DEFINITION. Let $\mathfrak{U} \in s(H)$. Then we say that A_H is $\mathfrak{U}$-*projective* if and only if $A_{\mathfrak{U},H} = A_H$.

If A_H has an identity element 1_H, then of course the condition $A_{\mathfrak{U},H} = A_H$ is equivalent to the condition $1_H \in A_{\mathfrak{U},H}$. To say that A_H is $\mathfrak{U}$-projective, is to say that A_H is "generated" by the algebras A_U, $U \in \mathfrak{U}$, in the sense that $A_H = \sum_{U \in \mathfrak{U}} A_U T_{U,H}$. We want next to define sets $\mathfrak{U}$ which are minimal with this property. It is convenient to take $H = G$ now, and to write j for j_G.

DEFINITION. $\mathfrak{D} \in s(G)$ is a *defect base* for A if and only if
(D1) $\mathfrak{D}$ is j-closed, and
(D2) For any j-closed $\mathfrak{U} \in s(G)$, A_G is $\mathfrak{U}$-projective $\Leftrightarrow \mathfrak{D} \subseteq \mathfrak{U}$.

THEOREM 1. *If A is any G-functor such that A_G has identity element 1_G, then A has a unique defect base $\mathfrak{D} = \mathfrak{D}(A)$.*

The uniqueness of $\mathfrak{D}$ is of course immediate from (D2); the existence of $\mathfrak{D}$ is proved by showing first that if $\mathfrak{U}, \mathfrak{V}$ are any j-closed elements of $s(G)$ such that $A_G = A_{\mathfrak{U},G} = A_{\mathfrak{V},G}$, then $\mathfrak{U} \cap \mathfrak{V}$ is j-closed, and $A_G = A_{\mathfrak{U} \cap \mathfrak{V}, G}$.

DEFINITION. A is a *local G-functor* if A_G is associative, has an identity element 1_G, and A_G is a *local ring*, i.e. if $A_G/J(A_G)$ is a division ring (here $J(A_G)$ is the Jacobson radical of A_G).

Suppose now that $\mathfrak{D}$ is the defect base of the local G-functor A. By (D2), $A_G = \sum A_{D,G}$, where D runs over $\mathfrak{D}$. Now because A_G is local, every *proper ideal* of A_G lies in $J(A_G)$. Since $J(A_G) < A_G$, not all the ideals $A_{D,G}$, $D \in \mathfrak{D}$, are proper; say $A_G = A_{D_0,G}$, for some $D_0 \in \mathfrak{D}$. It follows easily that $\mathfrak{D} = j\{D_0\}$.

DEFINITION. Let A be a local G-functor, and let D be a subgroup of G. Then we say D is a *defect group* of A if and only if $\mathfrak{D}(A) = j\{D\}$. This equation, it is easy to see, determines D up to conjugacy in G.

PROPOSITION. *Let A be a G-functor, such that A_H is associative with identity element 1_H, for all $H \leq G$. Let e be a local idempotent of A_G (i.e. e is an idempotent of A_G, and eA_Ge is a local ring). Then we may define a subfunctor $A' = eAe$ of A, by taking $A'_H = e_H . A_H . e_H$ ($e_H = eR_{G,H}$), for $H \leq G$. eAe is a local G-functor.*

DEFINITION. With the notation of the Proposition, and if D is a subgroup of G, we shall say D is a *defect group of e*, if and only if D is a defect group of the local G-functor eAe.

EXAMPLE 1. The best-known example of a defect base is that of the character ring functor A (this the G-functor for which A_H is the character ring of H). A_G is $\mathfrak{U}$-projective precisely when the unit character 1_G can be written as a sum of characters induced from characters on the subgroups $U \in \mathfrak{U}$. Brauer's famous "induction theorem" [2] shows that A_G is $\mathfrak{E}$-projective, where $\mathfrak{E}$ is the set of "elementary" subgroups of G. It is also easy to see that $\mathfrak{E}$ is minimal among j-closed $\mathfrak{U} \in s(G)$ for which A_G is $\mathfrak{U}$-projective, i.e. that $\mathfrak{D}(A) = \mathfrak{E}$.

We give next two examples, which come from G-algebras (Example 4). In each case, we suppose that k is an algebraically closed field of characteristic $p \neq 0$. We remark that if A_G is a finite-dimensional algebra over a field k, then an idempotent $e \in A_G$ is local in A_G, if and only if e is primitive in A_G.

EXAMPLE 4a. Take $A = kG$, the group algebra of G over k, and make it into a G-algebra by defining $\alpha^g = g^{-1}\alpha g$ ($\alpha \in A$, $g \in G$). Of course A_G is now the centre of A. Each local (= primitive) idempotent e_i of A_G corresponds uniquely to a p-block B_i of G in Brauer's sense. We find [4, p. 142] *D_i is a defect group of e_i in the G-functor defined by the G-algebra A if and only if D_i is a defect group of B_i in Brauer's definition* [1, p. 427].

EXAMPLE 4b. Let M be a finite-dimensional vector space over k, and let $\rho: G \to \text{End}_k(M)$ be a representation of G on M. Take $A = \text{End}_k(M)$ and make it into a G-algebra by defining $\alpha^g = \rho(g)^{-1}\alpha\rho(g)$ ($\alpha \in A$, $g \in G$). Then, for any $H \leq G$, $A_H = \text{End}_{kH}(M)$, and $A_{H,G} = A_G$ if and only if M (regarded as kG-module) is relatively (kG,kH)-projective. Now suppose that M is indecomposable (as kG-module). This is equivalent to saying that A_G is local (or that 1_G is a local idempotent in A_G).

D is a defect group of the local G-functor defined by the G-algebra $A = \text{End}_k(M)$, if and only if D is a vertex of M in the definition of [3].

4. **The transfer theorem.** Suppose now that A is any G-functor over k. In general, we should like to be able to describe the structure of the algebra A_G, in terms of the structure of the A_U, for some set of proper subgroups U of G. Although we are very far from achieving this, we have a "transfer theorem" (Theorem 2, below) which may give a little information of the sort we want, in suitable cases.

Let D, H be any subgroups of G such that $H \geq N_G(D)$. Define the following sets, which are both elements of $s(H)$:

$\mathfrak{X} = \{D \cap D^g | g \in G \backslash H\}$,
$\mathfrak{Y} = \{H \cap D^g | g \in G \backslash H\}$.

It is clear that $\mathfrak{X} \subseteq j_H\{D\}$, $\mathfrak{X} \subseteq j_H\mathfrak{Y}$, and hence

$$A_{\mathfrak{X},H} \leq A_{D,H} \cap A_{\mathfrak{Y},H} \ (=Q, \text{ say}).$$

Next we define four maps.

$t: A_{D,H}/A_{\mathfrak{X},H} \to A_{D,G}/A_{\mathfrak{X},G}.$
$r: A_{D,G}/A_{\mathfrak{X},G} \to A_{D,H} + A_{\mathfrak{Y},H}/A_{\mathfrak{Y},H}.$
$s: A_{D,H} + A_{\mathfrak{Y},H}/A_{\mathfrak{Y},H} \to A_{D,H}/Q.$
$q: A_{D,H}/A_{\mathfrak{X},H} \to A_{D,H}/Q.$

Here t is induced by $T_{H,G}: A_H \to A_G$, r by $R_{G,H}: A_G \to A_H$, s is the natural isomorphism, and q the natural epimorphism induced by the inclusion $A_{\mathfrak{X},H} \leq Q$ mentioned above. Of course, it is necessary to check that t,r are well defined.

THEOREM 2. *Let A be a G-functor over k, and D, H be subgroups of G such that $H \geq N_G(D)$. Define $\mathfrak{X}, \mathfrak{Y}, Q, t, s, r, q$ as above. Then*

(i) $trs = q$.

(ii) t,r *are both multiplicative and surjective.*

(iii) $\text{Ker } t \leq \text{Ker } q = Q/A_{\mathfrak{X},H}$. *Write $\bar{Q} = Q/A_{\mathfrak{X},H}$ and $\bar{A} = A_{D,H}/A_{\mathfrak{X},H}$. Then $\bar{Q}.\bar{A} = 0 = \bar{A}.\bar{Q}$. In particular this shows that $\text{Ker } t$ is nilpotent.*

(iv) $\text{Ker } r \leq \bar{Q}t$, *and $\bar{Q}t.\bar{A}t = 0 = \bar{A}t.\bar{Q}t$. In particular this shows that $\text{Ker } r$ is nilpotent.*

The condition $H \geq N_G(D)$ ensures that all the members of $\mathfrak{X}$ are *proper* subgroups of D; if this were not so, we should have $A_{D,H} = A_{\mathfrak{X},H}$ and our theorem would collapse. If $Q = A_{\mathfrak{X},H}$, then from (iii), (iv) we see that t,r,s,q are all isomorphisms. But even if $Q \neq A_{\mathfrak{X},H}$, then t,r,q are at least "near-isomorphisms," where a k-map $\theta: P \to P'$ is said to be a near-isomorphism if θ is multiplicative, surjective and $P.\text{Ker } \theta = \text{Ker } \theta.P = 0$.

A near-isomorphism θ induces a bijection from the set of idempotents of P onto the set of idempotents of P', as it is easy to show. For this reason, near-isomorphisms may be "as good as" isomorphisms for applications of Theorem 2.

One final remark: the map t is "filtration-preserving" in the following sense. Suppose $\mathfrak{Z} \in s(H)$ is such that

$$j_H\mathfrak{X} \subseteq j_H\mathfrak{Z} \subseteq j_H\{D\}.$$

Then $A_{\mathfrak{X},H} \leq A_{\mathfrak{Z},H} \leq A_{D,H}$, and $T_{H,G}$ maps these ideals of A_H onto the ideals $A_{\mathfrak{X},G} \leq A_{\mathfrak{Z},G} \leq A_{D,G}$ of A_G, and therefore t maps $A_{\mathfrak{Z},H}/A_{\mathfrak{X},H}$ onto $A_{\mathfrak{Z},G}/A_{\mathfrak{X},G}$, and $A_{D,H}/A_{\mathfrak{Z},H}$ onto $A_{D,G}/A_{\mathfrak{Z},G}$.

5. **An application of Theorem 2.** Theorem 2 is very general, and gives an almost bewildering variety of special cases. If we apply it to the cohomology ring functor (Example 2), taking $k = Z$ and looking only at the terms of degree 3, we get theorems identifiable as "transfer theorems" in the usual sense of finite group theory. This is because for any subgroup H of G, we can identify $\hat{H}^3(H,Z)$, $\hat{H}^3(G,Z)$ with the duals of H/H', G/G', and we can identify $T_{H,G}$ with the dual of Schur's

transfer $G/G' \to H/H'$. But it is not easy to recognise these transfer theorems in classical terms.

We shall give here one application of Theorem 2, namely to prove Brauer's "first main theorem on blocks" [1, p. 432]. We take the set-up of Example 4a, so that A is the G-functor defined by the G-algebra $A = kG$. Let D be a fixed p-subgroup of G, and H a subgroup of G such that $H \geqq N_G(D)$. Let $\{C_i\}$ be the set of all H-classes of G, i.e. the classes defined by the equivalence relation on G for which $g,g' \in G$ are equivalent if and only if $g' = h^{-1}gh$ for some $h \in H$. If C_i is such a class, write (C_i) for the sum of its elements in the group algebra $A = kG$. Then $\{(C_i)\}$ is a k-basis for A_H. Suppose now that $\mathfrak{U} \in s(H)$. By an easy calculation [4, p. 142] we find

(1) $A_{\mathfrak{U},H}$ has k-basis $\{(C_i)|\Delta_i \in j_H\mathfrak{U}\}$, where Δ_i denotes a *class-defect group of* C_i, i.e. Δ_i is a Sylow p-subgroup of $C_H(y_i)$, where y_i is some element of C_i. This definition fixes Δ_i only up to conjugacy in H. From (1) we find at once

(2) $Q = A_{D,H} \cap A_{\mathfrak{Y},H}$ is equal to $A_{\mathfrak{X},H}$. *Hence the maps t,r,s,q of Theorem 2 are all multiplicative isomorphisms.*

(3) $A_{D,H}/A_{\mathfrak{X},H}$ has k-basis $\{(C_i) + A_{\mathfrak{X},H}|\Delta_i \in j_H\{D\}\setminus j_H\mathfrak{X}\}$.

Next we observe that all the H-classes C_i which appear in (3) actually lie in H. For if C_i is such an H-class, we may assume that $\Delta_i \leqq D$ and $\Delta_i \notin j_H\mathfrak{X}$. Now Δ_i lies in $C_H(y)$ for some $y \in C_i$, hence $\Delta_i = \Delta_i^y$. But then $\Delta_i \leqq D^y$, so $\Delta_i \leqq D \cap D^y$, and if $y \notin H$, we should have the contradiction $\Delta_i \leqq D \cap D^y \in \mathfrak{X}$. Therefore y, and hence C_i, lies in H.

Now let $B = kH$, which we may regard as a subalgebra of A. B is an H-algebra, so we may define ideals $B_{D,H}, B_{\mathfrak{X},H}$ of B_H. By the same calculations as before we see that

(4) $B_{D,H}/B_{\mathfrak{X},H}$ has k-basis $\{(C_i) + B_{\mathfrak{X},H}|\Delta_i \in j_H\{D\}\setminus j_H\mathfrak{X}\}$. Of course the C_i in (4) are H-classes of H. But we have just proved that the C_i which appear in (3) are also in H. Hence

(5) *The inclusion $B \subseteq A$ induces an isomorphism*

$$m: B_{D,H}/B_{\mathfrak{X},H} \to A_{D,H}/A_{\mathfrak{X},H}.$$

Next we use Theorem 2, which gives us an isomorphism

$$t: A_{D,H}/A_{\mathfrak{X},H} \to A_{D,G}/A_{\mathfrak{X},G}.$$

Both m,t are "filtration-preserving," multiplicative isomorphisms, and therefore the same is true of their composite

$$mt: B_{D,H}/B_{\mathfrak{X},H} \to A_{D,G}/A_{\mathfrak{X},G}.$$

The p-blocks b_j of H correspond 1-1 with the local idempotents f_j of B_H ($=$ centre of B). Let D_j be a defect group of f_j, and let $\mathfrak{U} \in s(H)$. From our definition of defect group it is not hard to show

(6) $f_j \in B_{\mathfrak{U},H} \Leftrightarrow D_j \in j_H\mathfrak{U}$. In particular f_j lies in $B_{D,H}$ if and only if $D_j \in j_H\{D\}$. In this case, we shall assume (as we may, since D_j is determined only up to conjugacy in H) that $D_j \leqq D$. From (6) follows also

(7) *The local idempotents of $B_{D,H}/B_{\mathfrak{x},H}$ are those $f_j + B_{\mathfrak{x},H} = \bar{f}_j$ for which f_j have defect groups D_j such that $D_j \leq D$ and $D_j \notin j_H\mathfrak{X}$. Moreover for such an f_j, its defect group D_j is determined* (up to conjugacy in H) *as the smallest subgroup S of D such that $\bar{f}_j \in B_{\mathfrak{x}\cup\{S\},H}/B_{\mathfrak{x},H}$.*

The p-blocks B_i of G correspond 1-1 with the local idempotents e_i of A_G (= centre of A). We have at once a statement analogous to (7), which describes the local idempotents of $A_{D,G}/A_{\mathfrak{x},G}$. Applying now the filtration-preserving isomorphism mt, we deduce:

There is a 1-1 correspondence between the set of all blocks b_j of H having defect groups D_j such that $D_j \leq D$ and $D_j \notin j_H\mathfrak{X}$, and the set of all blocks B_i of G having defect groups D_i such that $D_i \leq D$ and $D_i \notin j_G\mathfrak{X}$.[1] If b_i, B_i are corresponding blocks of H, G respectively, then b_i has a defect group D_i which is also a defect group of B_i. In particular the set of all blocks of H with defect group D corresponds 1-1 with the set of all blocks of G with defect group D.

References

1. R. Brauer, *Zur Darstellungstheorie der Gruppen endlicher Ordnung*, Math. Z. **63** (1956), 406–444. MR **17**, 824.
2. R. Brauer and J. Tate, *On the characters of finite groups*, Ann. of Math. (2) **62** (1955), 1–7. MR **16**, 1087.
3. J. A. Green, *On the indecomposable representations of a finite group*, Math. Z. **70** (1958/59), 430–445. MR **24** #A1304.
4. ———, *Some remarks on defect groups*, Math. Z. **107** (1968), 133–150. MR **38** #2222.
5. ———, *Axiomatic representation theory for finite groups*, J. Pure Appl. Algebra. **1** (1971), 41–77.
6. T.-Y. Lam, *Induction theorems for Grothendieck groups and Whitehead groups of finite groups*, Ann. Sci. École Norm. Sup. (4) **1** (1968), 91–148. MR **38** #217.
7. T.-Y. Lam and I. Reiner, *Relative Grothendieck rings*, Bull. Amer. Math. Soc. **75** (1969), 496–498. MR **39** #326.

UNIVERSITY OF WARWICK, ENGLAND

---

[1] If D_j is a subgroup of D, the conditions $D_j \notin j_H\mathfrak{X}$ and $D_j \notin j_G\mathfrak{X}$ are equivalent.

Real Representations of Split Metacyclic Groups

Larry C. Grove

In [1], Brauer suggested the problem of determining how many irreducible complex representations of a finite group G can be written over the real field $\mathbf{R}$. An arithmetical answer for a class of metacyclic groups was given in [3]. The results of [3] are extended here to all split metacyclic groups, and hence, for example, to all groups having every Sylow subgroup cyclic.

Suppose then that
$$G = \langle a,b | a^m = b^s = 1, b^{-1}ab = a^r \rangle,$$

with $(m,r) = 1$ and $r^s \equiv 1 \pmod{m}$. We assume that $1 < r < m$, so G is not abelian. If u is the order of r modulo m, then $u|s$. We set $A = \langle a \rangle$ and $B = \langle b \rangle$, and denote by $\hat{A}$ the set $\{\varphi_0, \ldots, \varphi_{m-1}\}$ of irreducible C-characters of A, with $\varphi_0 = 1_A$. Then B acts, by conjugation, on $\hat{A}$ since $A \triangleleft G$. For each $\varphi_i \in \hat{A}$ we denote by $\mathcal{O}_i$ the B-orbit of φ_i and by B_i the stabilizer of φ_i in B, and we set $u_i = |\mathcal{O}_i|$.

If $\{\Psi_{ij} : 0 \leq j \leq s/u_i - 1\}$ are the characters of B_i, with $\Psi_{i0} = 1_{B_i}$, then the products
$$\chi_{ij} = \Psi_{ij}\varphi_i, \quad 0 \leq i \leq m-1, \quad 0 \leq j \leq s/u_i - 1,$$

are linear characters of $B_i A$. If one φ_i is chosen from each $\mathcal{O}_i$ and the resulting characters χ_{ij} are induced up to G then the set $\{\chi_{ij}^G\}$ is the set of distinct irreducible C-characters of G.

By the theorem of Frobenius and Schur [2, p. 21], χ_{ij}^G is the character of a real representation if and only if $v(\chi_{ij}^G) = |G|^{-1}\sum_{x \in G}\chi_{ij}^G(x^2) = 1$.

AMS 1970 *subject classifications*. Primary 20C15.

THEOREM 1. *If u_i is odd then $v(\chi_{ij}^G) = 0$ for all j, unless $i = 0$ or $i = m/2$. If $i = 0$ or $i = m/2$ then χ_{ij}^G is linear and $v(\chi_{ij}^G) = 1$ if and only if $j = 0$ or $j = s/2$.*

It follows that G has 1, 2, or 4 real linear characters, depending on the parities of m and s.

THEOREM 2. *If $i \neq 0$, $m/2$ then $v(\chi_{ij}^G) = 1$ if and only if $j = 0$, u_i is even, and $m|i(r^{u_i/2} + 1)$.*

If $0 < 2v|u$ we denote by M_v the set of divisors w of u maximal with respect to $w < 2v$. Then we set

$$d_v^{(0)} = d_v = (m, r^v + 1),$$

and

$$d_v^{(k)} = \sum \{(d_v, r^{(w_1, \ldots, w_k)} - 1) : w_i \in M_v, w_i \neq w_j \text{ if } i \neq j\},$$

for $1 \leq k \leq |M_v|$.

THEOREM 3. *The group G has exactly*

$$\sum_{2v|u} (2v)^{-1} \sum_{k \geq 0} (-1)^k d_v^{(k)}$$

distinct absolutely irreducible real representations.

COROLLARY. *A split metacyclic group G has all its representations real if and only if $G = D_m$, the dihedral group of order $2m$.*

Details will appear elsewhere.

REFERENCES

1. R. Brauer, *Representations of finite groups*, Lectures on Modern Mathematics, vol. 1, Wiley, New York, 1963, pp. 133–175. MR **31** #2314.
2. W. Feit, *Characters of finite groups*, Benjamin, New York, 1967. MR **36** #2715.
3. L. C. Grove, *Real representations of metacyclic groups*, Proc. Amer. Math. Soc. **21** (1969), 417–421. MR **38** #4571.

SYRACUSE UNIVERSITY

On Some Doubly Transitive Groups

Koichiro Harada[1]

1. **Introduction.** We consider a finite group G satisfying the following condition (∗).

(∗) $\begin{cases} (1) \ G \text{ is a doubly transitive group on } \Omega = \{1,2,\ldots,n\}, \\ (2) \text{ if } K \text{ is the stabilizer of two different points, then } K \text{ has even order, and} \\ (3) \ K \cap K^g \text{ has odd order if } g \in G \text{ does not normalize } K. \end{cases}$

THEOREM 1. *Let G be a group satisfying* (∗). *If the degree n of G is odd, then*

(1) *G has a regular normal subgroup R and an involution z such that $G = C_G(z) \cdot R$,*

or

(2) *a Sylow 2-subgroup of G is dihedral, quasi-dihedral, wreathed product $Z_{2^n} \wr Z_2$ or $Z_{2^n} \times Z_{2^n}$, $n \geq 2$.*

Theorem 1 is an easy consequence of the following more general result which has its own interest.

THEOREM 2. *Let G be a finite group and H be a proper subgroup of even order. Assume*

(1) *$H \cap H^g$ has odd order if $g \notin N_G(H)$,*

(2) *a Sylow 2-subgroup of $N_G(H)/H$ is cyclic or generalized quaternion, and*

(3) *$N_G(H) - H$ contains an element which is conjugate to a 2-element of H in G.*

Then one of the following conditions holds:

(1) *G has a normal subgroup N of index 2 such that $N \cap H$ is of odd order,*

(2) *G contains an involution z such that $G = C_G(z) \cdot O(G)$,*

(3) *a Sylow 2-subgroup of G is dihedral, quasi-dihedral, wreathed product $Z_{2^n} \wr Z_2$ or $Z_{2^n} \times Z_{2^n}$, $n \geq 2$.*

AMS 1970 *subject classifications*. Primary 20B20; Secondary 20F25, 20D20, 20G40.

[1] This research was supported in part by National Science Foundation grant GP-7952X.

EXAMPLE. Assume $G = SL(3, q)$, q odd. Let L be the centralizer of an involution of G. Then $L \cong GL(2,q)$. Let H be a subgroup L such that $H \cong SL(2,q)$. Then, G, H satisfy all the hypotheses of Theorem 2.

THEOREM 3 (C. HERING [6] AND J. KING [10]). *Let G be a group satisfying (∗). Suppose that the number of fixed points by K is two or three, then a Sylow 2-subgroup of G is dihedral or quasi-dihedral.*

Hence by [1], [5], the structure of G is known.

THEOREM 4. *Let G be a group satisfying (∗). If the degree n of G is even, then*

(1) *G is isomorphic to a subgroup of $P\Gamma L(2,q_1)$ containing $PSL(2,q_1)$ where q_1 is odd and $n = q_1 + 1$, or*

(2) *G is isomorphic to an automorphism group of $AG(d,q_2)$ where $d = 1,2$ and q_2 is even, or*

(3) *$G \cong A_6$ and $n = 6$, or $G \cong P\Gamma L(2,8)$ and $n = 28$.*

Let G be a permutation group on $\Omega = \{1, 2, \ldots, n\}$. Denote by $G_{1,2,\ldots}$ the set of elements of G which fix $1, 2, \ldots$, point-wise. For a subgroup H of G, $I(H)$ denotes the set of all points $\in \Omega$ which are fixed by every element of H. We remark here that the condition (3) of (∗) implies that if the set $\{1, 2, \ldots\}$ has $|I(K) + 1|$ points, then $G_{1,2,\ldots}$ has odd order. Equivalently, (3) implies that a Sylow 2-subgroup of K acts semiregularly on $\Omega - I(K)$.

2. **Proof of Theorem 2.** Assume that G and H satisfy the conditions of Theorem 2. Let S be a Sylow 2-subgroup of $N_G(H)$. Then $T = S \cap H$ is a Sylow 2-subgroup of H. By assumption there exists an element y of $N_G(H) - H$ which is conjugate to a 2-element x of H. Put $y = x^g$, $g \in G$. We fix G, H, S, T, x, y, g throughout this section.

First we note that the condition (1) implies that $N_G(H)$ contains the normalizer in G of every nontrivial group of even order of H. We have two cases:

[I] *S is not a Sylow 2-subgroup of G, or*

[II] *S is a Sylow 2-subgroup of G.*

Assume that [I] is the case. Then G contains a 2-element $a \in N_G(S) - S$. By our assumption, $T \cap T^a = 1$. Hence T is isomorphic to a subgroup of S/T, as $T^a \subset S$ and $T \cong T^a$. Hence T is cyclic or generalized quaternion. This, together with the fact that $N_G(H)$ contains the centralizer of every 2-element $\neq 1$ of H, restricts the structure of S very much. Applying either Thompson's transfer lemma or Glauberman's Z^*-theorem, we can claim:

PROPOSITION. *If S is not a Sylow 2-subgroup of G, then one of the following conditions hold:*

(1) *$G = C_G(z) \cdot O(G)$ for some involution z which is not conjugate to an involution of H,*

(2) *G contains a normal subgroup N of index 2 such that $N \cap H$ is of odd order, or*

(3) *a Sylow 2-subgroup of G is dihedral or quasi-dihedral.*

In the proof of this proposition and in the rest of this note we frequently use the well-known fact that if a 2-group X does not contain a normal subgroup of type (2,2), then X is cyclic, or of maximal class.

Next assume that S is a Sylow 2-subgroup of G. We reduce the proof to the case that $|T| \geqq 4$, y is an involution, $x \in Z(S) \cap T$, $|C_T(y)| \geqq 4$ and S is neither dihedral nor quasi-dihedral. We can claim:

LEMMA. *S contains unique normal subgroup U of type (2,2). Furthermore $N_G(U)/C_G(U) \cong S_3$.*

LEMMA. *$C_S(U)/C_T(U)$ and $C_T(U)$ are cyclic or generalized quaternion.*

Hence in order to prove that $S \cong Z_{2^n} \times Z_{2^n}$ or $S \cong Z_{2^n} \wr Z_2$, it is enough to show:

LEMMA. *Let X be a 2-group and Y a nontrivial normal subgroup of X. Assume that X/Y and X are cyclic or generalized quaternion, that $\mathrm{Aut}(X)$ contains α of odd order which does not centralize an (unique) involution of Y. Then $X \cong Z_{2^n} \times Z_{2^n}$.*

The proof of this lemma is elementary.

3. **Proof of Theorem 1 and Theorem 3.** Setting $K = G_{1,2}$, we easily see that G and K satisfy all the conditions of Theorem 2 if we replace H by K. If (1) of the conclusion of Theorem 2 occurs, we can claim that N is doubly transitive on Ω. Using [3], we have our theorem in this case as well.

4. **Proof of Theorem 4.** We set $H = G_1$ and $K = G_{1,2}$. Since n is even, $|I(K)|$ is even. Hence $H_G(K)/K$ is a frobenius group of order $q(q-1)$, $q = 2^m$. The following is an easy consequence of our assumption.

LEMMA. *$N_H(K)$ is a strongly embedded subgroup of H.*

Hence by [4], we have two cases to consider:

[I] *A Sylow 2-subgroup of H is cyclic or generalized quaternion, or*

[II] *$H/O(H)$ contains a normal subgroup of odd index isomorphic to $SL(2, 2^e)$, $Sz(2^e)$ or $PSU(3, 2^e)$.*

Now suppose that [I] is the case. Then

LEMMA. *A Sylow 2-subgroup of K is cyclic of order 2.*

LEMMA. *A Sylow 2-subgroup of $N_G(K)$ is elementary of order $2q$.*

LEMMA. *If G has single conjugacy classes of involutions, then $G \cong P\Gamma L(2,8)$.*

In order to prove the last lemma, we require [8]. Thus we assume that G has at least two conjugacy classes of involutions. We claim:

LEMMA. *(i) G has exactly two conjugacy classes of involutions. (ii) $N_G(S)/C_G(S)$ is a frobenius group of order $(q-1)q$ where S is a Sylow 2-subgroup of $N_G(K)$.*

By this lemma, we know the structure of a Sylow 2-subgroup S_1 of $N_G(S)$. If S_1 is a Sylow 2-subgroup of G, we can claim that G has a normal subgroup of index 2 which does not contain an involution of K. By [2] and [7] we have the desired result. If S_1 is not a Sylow 2-subgroup of G, then we prove

LEMMA. $N_G(S_1)/C_G(S_1) \cdot S_1$ is a frobenius group of order $(q-1)q$.

Continuing this "pushing up" method, we finally reach a contradiction.

Next assume that [II] is the case. Consider the set $\Delta = \{X^h | h \in H, X = I(K)\}$. Then H induces a permutation group $\bar{H}$ on Δ. We can claim:

LEMMA. $\bar{H}$ contains a normal subgroup of odd index which is isomorphic to $SL(2, 2^e)$, $Sz(2^e)$ or $PSU(3, 2^e)$ with their natural permutation representations.

This lemma enables us to quote [9] and to obtain the desired conclusion in this case as well.

REFERENCES

1. J. Alperin, R. Brauer, and D. Gorenstein, *Finite groups with quasi-dihedral and wreathed Sylow 2-subgroups*, Trans. Amer. Math. Soc. **151** (1970), 1–261.

2. H. Bender, *Endliche zweifach transitive Permutationsgruppen, deren Involutionen kleine Fixpunkte haben*, Math. Z. **104** (1968), 175–204. MR **37** #2846.

3. ———, *Doubly transitive groups with no involution fixing two points* (to appear).

4. ———, *Finite groups having a strongly embedded subgroup* (to appear).

5. D. Gorenstein and J. H. Walter, *The characterization of finite groups with dihedral Sylow 2-subgroups*. I–III, J. Algebra **2** (1965), 85–151, 218–270, 354–393. MR **31** #1297a,b; MR **32** #7634.

6. C. Hering, *Zweifach transitive Permutationsgruppen, in denen 2 die maximale Anzahl von Fixpunkten von Involutionen ist*, Math. Z. **104** (1968), 150–174. MR **37** #295.

7. B. Huppert, *Zweifach transitive, auflösbare Permutationsgruppen*, Math. Z. **68** (1957), 126–150. MR **20** #904.

8. N. Ito, *On doubly transitive groups of degree n and order $2(n-1)n$*, Nagoya Math. J. **27** (1966), 409–417. MR **34** #247.

9. W. Kantor, *On 2-transitive groups in which the stabilizer of two points fixes additional points* (to appear).

10. J. D. King, *A characterization of some doubly transitive groups*, Math. Z. **107** (1968), 43–48. MR **38** #3329.

THE INSTITUTE FOR ADVANCED STUDY

Characterization of Rank 3 Permutation Groups by the Subdegrees

D. G. Higman

A procedure which is sometimes effective in characterizing a class of groups involves three steps: (1) the determination of faithful representations of the groups in the automorphism groups of some class of combinatorial configurations, (2) the determination of the relevant configurations, and (3) the determination of the relevant subgroups of the automorphism groups of these configurations. In studying rank 3 permutation groups of even order, we can start from the fact that such groups are represented in the automorphism groups of strongly regular graphs ([1] is a convenient self-contained reference for basic facts and notation for rank 3 groups). The following theorem is proved in [2] by carrying out the indicated procedure. We write $Q_m = q^{m-1} + q^{m-2} + \cdots + q + 1$ and $Q_{m,2} = Q_m Q_{m-1}/Q_2$. Thus if q is a prime power ≥ 2 and m is an integer ≥ 2, Q_m and $Q_{m,2}$ are respectively the number of points and the number of lines in projective space $P_{m-1}(q)$ of dimension $m - 1$ over the field q elements.

THEOREM. *Let G be a rank 3 permutation group of degree $n = Q_{m,2}$ with subdegrees $k = qQ_2Q_{m-2}$ and $l = q^4 Q_{m-2,2}$, where $q \geq 1$ and $m \geq 4$ are integers. Then*

I. *For $q = 1$, one of the following holds:*

(a) *G is isomorphic with a 4-fold transitive group of degree m in its action on the 2-element subsets,*

(b) *G is isomorphic with $P\Gamma L_2(8)$ in its action on the 2-element subsets of the 9 points of the projective line,*

(c) *$\mu = 6$ and $m = 9, 17, 27$ or 57,*

AMS subject classifications. Primary 20B10, 20B25; Secondary 05B25.

(d) $\mu = 7$ and $m = 51$,
(e) $\mu = 8$, and $m = 28, 36, 325, 903$, or $8,128$.

II. *For $q \geq 2$, one of the following holds:*

(a) *G is isomorphic with a subgroup H of $P\Gamma L_m(q)$ in its action on the lines of $\boldsymbol{P}_{m-1}(q)$, and $H \geq PSL_m(q)$ or $q = 2$ and H is transitive on the 4-simplices,*

(b) $m = 4$ *or* 5,

(c) *m is odd, $7 \leq m \leq 17$ and $\mu \neq (q+1)^2$.*

In I, the case $\mu = 6$ and $m = 9$ is realized by the automorphism group of $G_2(2)$, but existence in the remaining cases in (c), (d) and (e) is undecided. As far as we know, the conclusion (a) in II may hold for all $q \geq 2$ and $m \geq 4$. Whether or not $H \geq PSL_m(2)$ in (a) of II in case $q = 2$ is undecided.

REFERENCES

1. M. D. Hestenes and D. G. Higman, *Rank 3 and strongly regular graphs*, SIAM-AMS Proc. (to appear).

2. D. G. Higman, *Characterizations of families of rank 3 permutation groups by the subdegrees*, I, II, Arch. Math. **21** (1970), 151–156; II (to appear).

UNIVERSITY OF MICHIGAN

Symplectic Action and the Schur Index

I. M. Isaacs

Let K be a finite field of characteristic p and let V be a vectorspace of finite dimension over K. Suppose that V is equipped with a nondegenerate alternating form and that a group, G, of p'-order acts on V so as to preserve the form. For $g \in G$, it is not hard to see that $C_V(g)$ is a subspace on which the form is nondegenerate and thus $\dim_K C_V(g)$ is even. In other words, $|C_V(g)|$ is a square and the permutation character, π, of G on the elements of V takes on square values.

THEOREM. *There exists a sign function, s, on G with $s(g) = \pm 1$ such that*

$$\psi(g) = s(g)\sqrt{\pi(g)}$$

is an ordinary, integral valued (reducible) character of G, afforded by a representation over $\mathbf{Q}[\eta]$ where η is a primitive pth root of unity if $p \neq 2$ and a primitive 4th root if $p = 2$.

The following recipe may be used to calculate such a sign function explicitly. It works by double induction on $|G|$ and $|V|$.

(1) $s(1) = +1$.

(2) If $H \subseteq G$ and $h \in H$, then $s_H(h) = s(h)$, where s_H is the sign function associated with the action of H on V.

(3) Let $N = C_G(V)$. Then $\bar{G} = G/N$ acts on V and for $g \in G$ we have $s(g) = s_{\bar{G}}(\bar{g})$ where $\bar{g}$ is the image of g in $\bar{G}$.

Because of (2), we may assume that $G = \langle g \rangle$ and by (3), we may assume that G acts faithfully on V.

(4) Suppose there exists $x \in G$ with $(0) < C_V(x) < V$. Then $V = V_1 \dotplus V_2$ where

---

AMS 1970 *subject classifications.* Primary 20C15.

$V_1 = C_V(x)$ and $V_2 = \{v^x - v | v \in V\}$. The form is nondegenerate on both V_1 and V_2 and G acts on each of these spaces yielding sign functions s_1 and s_2. In this case, $s(g) = s_1(g)s_2(g)$.

We may now assume that G acts in a fixed-point-free (Frobenius) manner on V.

(5) Put $|V| = e^2, n = o(g) = |G|$. Then $n|(e \pm 1)$. (Note that $n|(e^2 - 1)$ is clear but the stronger statement is true also.) Set $\varepsilon = \pm 1$ so that $n|(e - \varepsilon)$. If $n = 2$, choose ε so that $4|(e - \varepsilon)$. Now

(a) if n is odd, then $s(g) = \varepsilon$;
(b) if n is even and $(e - \varepsilon)/n$ is even, then $s(g) = \varepsilon$;
(c) if n is even and $(e - \varepsilon)/n$ is odd, then $s(g) = -\varepsilon$.

In case G is a q-group for an odd prime q, an easier determination of the signs is given by the formula

$$\psi(g) \equiv \sqrt{|V|} \bmod q.$$

As an example of how the above can be applied, we calculate some Schur indices for generalized quaternion groups.

PROPOSITION. *Let $G = Q_{2^{n+1}}$ with $n \geq 2$ and let p be a prime with $p \equiv 2^n \pm 1$ mod 2^{n+1}. Let η be a primitive pth root of unity and $F = \mathbf{Q}[\eta]$. Then $m_F(\chi) = 1$ for all irreducible characters, χ, of G.*

COROLLARY. *Let $p \equiv 3$ or 5 mod 8 be a prime, and let η be a primitive pth root of unity. Then -1 is a sum of two squares in $\mathbf{Q}[\eta]$.*

PROOF OF PROPOSITION. We have $G \subseteq \mathrm{SL}(2, p) = \mathrm{Sp}(2, p)$. Let $K = \mathrm{GF}(p)$ and $\dim_K(V) = 2$. Then G acts on V Frobeniusly and preserves a nondegenerate alternating form. Now $\pi(1) = p^2$ and $\pi(g) = 1$ for $g \in G^\#$. Thus $\psi(1) = p$ and $\psi(g) = s(g)$ for $g \in G^\#$ defines a character of G afforded by an F-representation. In order to show that $m_F(\chi) = 1$, it suffices to show that (χ, ψ) is odd whenever $\chi(1) = 2$.

To calculate ψ, define $\delta = \pm 1$ so that $p \equiv 2^n + \delta \bmod 2^{n+1}$. Thus $2^n|(p - \delta)$ and in the notation of rule (5), $e = p$ and $\varepsilon = \delta$. If $g \in G$ is of order 2^a, with $a > 0$, we have:

(a) if $a < n$, then $(e - \varepsilon)/2^a$ is even and $\psi(g) = s(g) = \delta$;
(b) if $a = n$, then $(e = \varepsilon)/2^a$ is odd and $\psi(g) = s(g) = -\delta$.

Let $Z \subseteq G$ be cyclic of order 2^n. It follows that

$$\psi_Z = \delta\mu + \frac{p - \delta}{2^n}\rho,$$

where ρ is the regular character of Z and μ is the unique nonprincipal linear character with $\mu^2 = 1_Z$. If $\chi(1) = 2$, then $\chi = \lambda^G$ for a linear character $\lambda \neq \mu$ of Z. Then $(\chi, \psi) = (\lambda, \psi_Z) = (p - \delta)/2^n$ is odd and the proof is complete.

The proofs of the properties of ψ and of the rules for calculating the signs are implicit in §5 of [3]. In order to convert the present situation into the one described there, one constructs a p-group, H, admitting G, with $H/Z(H)$ and V isomorphic as G-modules. This is done so that H has an irreducible G-invariant character θ of degree $\sqrt{|V|}$. Then ψ is the restriction to G of a suitable extension of θ to HG. For odd p, H is an extra-special group of exponent p. If $p = 2$, H is the central product of an extra-special group and a cyclic group of order 4.

Results similar to those in §5 of [3] have recently appeared in [1]. A much sharper version of the corollary given here about -1 as a sum of two squares in cyclotomic fields is proved in [2]. The proof there is not group theoretic.

REFERENCES

1. T. R. Berger, *Class two p-groups as fixed point free automorphism groups*, Illinois J. Math. **14** (1970), 121–149.

2. B. Fein and B. Gordon, *On the representation of -1 as a sum of two squares in an algebraic number field*, J. Numer. Theory (to appear).

3. I. M. Isaacs, *Extensions of group representations over nonalgebraically closed fields*, Trans. Amer. Math. Soc. **141** (1969), 211–228. MR **39** #4297.

UNIVERSITY OF WISCONSIN

On Factorizable Groups

Noboru Ito[1]

In the present note we introduce two families of finite groups which are related in a natural way (§1). Then we raise a fundamental question concerning these families of groups. Further, we state some propositions which answer affirmatively the question in special cases, together with historical remarks (§2). In the last section (§3) we give proofs to two of the propositions stated in §2.

NOTATION. Let $\mathfrak{G}$ be a finite group. Let $\mathfrak{X}$ be a subset of $\mathfrak{G}$. $Cs\mathfrak{X}$ and $Ns\mathfrak{X}$ denote the centralizer and normalizer of $\mathfrak{X}$ in $\mathfrak{G}$. $|\mathfrak{X}|$ denotes the number of elements in $\mathfrak{X}$. $\langle \mathfrak{X} \rangle$ denotes the subgroup of $\mathfrak{G}$ generated by $\mathfrak{X}$. Let $\mathfrak{H}$ be a subgroup of $\mathfrak{G}$. Core $\mathfrak{H}$ denotes the largest normal subgroup of $\mathfrak{G}$ contained in $\mathfrak{H}$. $\mathfrak{G}^*$ denotes the least normal subgroup of $\mathfrak{G}$ such that $\mathfrak{G}/\mathfrak{G}^*$ is solvable. $Z(\mathfrak{G})$ denotes the center of $\mathfrak{G}$. $GL(r, q)$ denotes the r-dimensional general linear group over the field of q elements. $V(r, q)$ denotes the r-dimensional vector space over the field of q elements. A_n denotes the alternating group of degree n.

1.

DEFINITION 1. A finite group $\mathfrak{G}$ belongs to the family $\mathscr{C}$ if and only if there exist two elements $A \neq E$ and $B \neq E$ of $\mathfrak{G}$ such that $\mathfrak{G} = CsA \cdot CsB$.

The family $\mathscr{C}$ is huge. In fact, if $\mathfrak{A} \neq \mathfrak{E}$ and $\mathfrak{B} \neq \mathfrak{E}$ are any two groups, then their direct product $\mathfrak{G} = \mathfrak{A} \times \mathfrak{B}$ belongs to $\mathscr{C}$.

DEFINITION 2. A group $\mathfrak{G}$ belongs to the family $\mathscr{X}$ if and only if there exist two conjugacy classes $[A] \neq [E]$ and $[B] \neq [E]$ such that $[A][B]$ is a scalar multiple of the conjugacy class $[AB]$:

(1) $$[A][B] = m[AB].$$

AMS 1970 *subject classifications.* Primary 20D40, 20D05; Secondary 20B20.

[1]This work is supported in part by NSF Grant GP 9584.

PROPOSITION 1. *If* $\mathfrak{G} = CsA \cdot CsB$, *then* $[A][B] = m[AB]$ *with* $m = |Cs(AB)|/|CsA \cap CsB|$. *If* $[A][B] = m[AB]$, *then* $m \geq |Cs(AB)|/|CsA \cap CsB|$. *In particular, if* $m = |Cs(AB)|/|CsA \cap CsB|$, *then* $\mathfrak{G} = CsA \cdot CsB$.

PROOF. Let X and Y be any two elements of $\mathfrak{G}$. Write $X = X_1 X_2$ with $X_1 \in CsA$ and $X_2 \in CsB$. Since $\mathfrak{G} = CsX_2^{-1}BX_2 \cdot CsX_2^{-1}AX_2$, write $Y = Y_1 Y_2$ with $Y_1 \in CsX_2^{-1}BX_2$ and $Y_2 \in CsX_2^{-1}AX_2$. Then we have that

$$X^{-1}AXY^{-1}BY = X_2^{-1}AX_2Y^{-1}BY = X_2^{-1}AX_2Y^{-1}X_2^{-1}BX_2Y$$
$$= X_2^{-1}AX_2Y_2^{-1}X_2^{-1}BX_2Y_2 = Y_2^{-1}X_2^{-1}AX_2Y_2Y_2^{-1}X_2^{-1}BX_2Y_2$$
$$= Y_2^{-1}X_2^{-1}ABX_2Y_2.$$

$[A]$, $[B]$, and $[AB]$ contain $|\mathfrak{G}|/|CsA|$, $|\mathfrak{G}|/|CsB|$ and $|\mathfrak{G}|/|CsAB|$ elements. Hence $m = |\mathfrak{G}||CsAB|/|CsA||CsB|$. Since $|\mathfrak{G}| \geq |CsA \cdot CsB| = |CsA||CsB|/|CsA \cap CsB|$, we obtain that $m \geq |CsAB|/|CsA||CsB|$ and that the equality holds if and only if $\mathfrak{G} = CsA \cdot CsB$.

REMARK 1. Proposition 1 shows that $\mathscr{C} \geq \mathscr{X}$. Now let $\mathfrak{H}$ be a subgroup of GL(4,2) which is isomorphic to A_7. Let $\mathfrak{G}$ be the split extension of $V(4,2)$ by $\mathfrak{H}$. Then it is not difficult to verify that $\mathfrak{G}$ belongs to $\mathscr{X}$ and that $\mathfrak{G}$ does not belong to $\mathscr{C}$.

PROPOSITION 2. $[A][B] = m[AB]$ *if and only if*

(2) $$\chi(A)\chi(B) = \chi(E)\chi(AB)$$

for every irreducible character χ *of* $\mathfrak{G}$.

PROOF. It follows from $[A][B] = m[AB]$ that $|\mathfrak{G}|\chi(A)|\mathfrak{G}|\chi(B)/|CsA|\chi(E)|CsB|\chi(E) = m|\mathfrak{G}|\chi(AB)/|CsAB|\chi(E)$.

Since $m = |\mathfrak{G}||CsAB|/|CsA||CsB|$, we have that $\chi(A)\chi(B) = \chi(E)\chi(AB)$.

On the other hand, assume that (2) holds. Put $[A][B] = m[AB] + \mathfrak{X}$, where $\mathfrak{X}$ denotes a linear combination of conjugacy classes of G other than $[AB]$. Then by (1) we obtain that

$$m = (|\mathfrak{G}|/|CsA||CsB|) \sum_\chi \chi(A)\chi(B)\overline{\chi(AB)}/\chi(E)$$
$$= (|\mathfrak{G}|/|CsA||CsB|) \sum_\chi \chi(AB)\overline{\chi(AB)}$$
$$= |\mathfrak{G}||CsAB|/|CsA||CsB|.$$

Hence we have that $|\mathfrak{X}| = 0$ and that $[A][B] = m[AB]$.

2.

QUESTION. Let $\mathfrak{G}$ be a group in the family $\mathscr{C}$ (or $\mathscr{X}$). If $\mathfrak{G}$ is simple, then is $\mathfrak{G}$ a cyclic group of a prime order?

In the case of the family $\mathscr{X}$ we failed to find a result on this question in the literature. We can mention the following result. But we omit the proof, because our proof is rather lengthy.

PROPOSITION 3. *If $\mathfrak{G}$ belongs to the family $\mathscr{X}$ and if $\mathfrak{G}$ is a quadruply transitive permutation group, then $\mathfrak{G}$ is the symmetric group of degree* 4.

REMARK 2. If we assume furthermore that $\mathfrak{G}$ belongs to the family $\mathscr{C}$, then Proposition 3 can be easily proved.

On the other hand, in the case of the family $\mathscr{C}$ we can find many results in literature:

(i) (Burnside, [3, (18.2)].) If $\mathfrak{G}:CsA$ is a power of a prime p, then $\mathfrak{G}$ is not simple.

We notice that $\mathfrak{G}$ belongs to $\mathscr{C}$. In fact, let $B \neq E$ be an element of the center of a Sylow p-subgroup of $\mathfrak{G}$. Then we have obviously that $\mathfrak{G} = CsA \cdot CsB$.

(ii) (Wielandt-Kegel, [7, p. 674].) If $\mathfrak{G}$ is a product of two nilpotent subgroups, then $\mathfrak{G}$ is solvable.

(iii) (Szép, [10], [11].) Let $\mathfrak{G} = CsA \cdot \mathfrak{B}$ with $\mathfrak{B}$ abelian. If either $|Z(CsA)| \geq |\mathfrak{B}|$ or $|\mathfrak{B}| \geq |CsA|$, then $\mathfrak{G}$ is not simple.

(iv) (Fischer [5], Ito [8], Wielandt[2].) Let $\mathfrak{G}:CsA = p$ (p: a prime) and let core $(CsA) = \mathfrak{E}$. Then $\mathfrak{G}$ is solvable.

The following proposition is trivial, but useful.

PROPOSITION 4. *Let $\mathfrak{G} = CsA \cdot CsB$. If $AB = BA$, then $\mathfrak{G}$ is not simple. If $m = 1$ in* (1), *then $AB = BA$.*

PROOF. If $AB = BA$, then core (CsA) contains B. If $m = 1$, then $CsAB = CsA \cap CsB$. Hence $AB = BA$.

Now the main purpose of this paper is to prove the following two propositions, which also answer the above question affirmatively in special cases.

PROPOSITION 5. *Let $\mathfrak{G} = CsA \cdot CsB$. If $\mathfrak{G}:CsA = pq$ with p and q primes, then $\mathfrak{G}$ is not simple.*

REMARK 3. We notice that $GL(3,2) = \mathfrak{A}\mathfrak{B}$, where $\mathfrak{A}$ is a Sylow 2-subgroup of index $21 = 3.7$ and $\mathfrak{B}$ is a metacyclic group of order 21.

Assume that core $(CsA) = \mathfrak{E}$ in Proposition 5. If $p = q$, we cannot claim that G is solvable. In fact, let $p \geq 5$ and let $\mathfrak{G}$ be the split extension of $V(2,p)$ by $GL(2,p)$. Then it is easy to verify that $\mathfrak{G}$ belongs to $\mathscr{C}$ and that every central element $\neq E$ of $GL(2,p)$ is of index p^2 in $\mathfrak{G}$.

PROPOSITION 6. *Assume that core $(CsA) = \mathfrak{E}$ and that $p \neq q$ in Proposition 5. Then $\mathfrak{G}$ is solvable.*

Finally we mention without proof the following proposition.

PROPOSITION 7. *Let $\mathfrak{G} = CsA \cdot CsB$. If $\mathfrak{G}$ possesses an irreducible character χ of a prime degree p, then $\mathfrak{G}$ is not simple.*

---

[2]Unpublished.

3.

PROOF OF PROPOSITION 5. (a) Assume that $\mathfrak{G}$ is simple. By a theorem of Burnside in §2 we may assume that $p \neq q$. Further we may assume that $|\langle B \rangle| = r$ is a prime. If $r \neq p, q$, then, since CsA contains a Sylow r-subgroup of $\mathfrak{G}$, we may assume that B belongs to CsA. Then core (CsA) contains B. Hence we may assume that $|\langle B \rangle| = p$ and that $AB \neq BA$. Since a Sylow q-subgroup of CsB is not contained in CsA, there exists a q-element C of CsB which does not belong to CsA. Therefore $\mathfrak{G} = CsA \cdot \langle BC \rangle$. If $|\langle C \rangle| > q$, then C^q belongs to core (CsA). Hence we may assume that $|\langle C \rangle| = q$. Therefore $\mathfrak{G} = CsA \cdot \langle BC \rangle$ and $CsA \cap \langle BC \rangle = \mathfrak{E}$. We may assume that CsA is a maximal subgroup of $\mathfrak{G}$. So by a theorem of Schur [**12**, (25.3)] the permutation representation of $\mathfrak{G}$ induced by CsA is doubly transitive. We may assume that $\mathfrak{G}$ is a doubly transitive permutation group on $\Omega = \{1, 2, \ldots, pq\}$ and that CsA is the stabilizer of the point 1 in $\mathfrak{G}$.

(b) Since $BC = CB$, we may assume that $p > q$. Let $\mathfrak{P}_1$ be a Sylow p-subgroup of CsA such that $\mathfrak{P} = \mathfrak{P}_1 \cdot \langle B \rangle$ is a Sylow p-subgroup of $\mathfrak{G}$. Since $p > q$, $\mathfrak{P}$ is elementary abelian. We show that $\mathfrak{P}_1 = \mathfrak{E}$. Assume that $\mathfrak{P}_1 \neq \mathfrak{E}$. Let $B = \pi_1 \ldots \pi_q$ be the cycle structure of B. Then CsB can be represented as a permutation group on $\{\pi_1, \ldots, \pi_q\}$. The kernel of this representation is a p-group. Therefore it coincides with $\mathfrak{P} = \mathfrak{P}_1 \cdot \langle B \rangle$. So $\mathfrak{P} \cdot \langle C \rangle$ is a transitive subgroup of $\mathfrak{G}$. Further $\mathfrak{P}$ is normal in $\mathfrak{P} \cdot \langle C \rangle$. Let $\alpha(X)$ be the set of fixed points of a subset $\mathfrak{X}$ of $\mathfrak{G}$ in Ω. Then by a theorem of Jordan [**12**, (3.5)] $Ns\mathfrak{P}_1 \cap \mathfrak{P}\langle C \rangle$ is transitive on $\alpha(\mathfrak{P}_1)$. If $Ns\mathfrak{P}_1$ contains C, then core (CsA) contains $\mathfrak{P}_1$ against the assumption. So we have that $Ns\mathfrak{P}_1 \cap \mathfrak{P}\langle C \rangle = \mathfrak{P}$. Since $|\langle B \rangle| = p$, we have that $|\alpha(\mathfrak{P}_1)| = p$. Now by a theorem of Witt [**12**, (9.4)] $Ns\mathfrak{P}_1$ is doubly transitive on $\alpha(\mathfrak{P}_1)$. A belongs to the center of the stabilizer of the point 1 in $Ns\mathfrak{P}_1$. Therefore by (iv) in §2 $Ns\mathfrak{P}_1/\mathfrak{N}$ is solvable of order $p(p-1)$, where $\mathfrak{N}$ is the kernel of the permutation representation of $Ns\mathfrak{P}_1$ on $\alpha(\mathfrak{P}_1)$. Now $Cs\mathfrak{P}_1$ contains $\mathfrak{P}$ and A. Put $\mathfrak{Z}^{\#} = \mathfrak{P} \cap Z(Ns\mathfrak{P}) \cap Cs\mathfrak{P}_1$. By a theorem of Turkin-Zassenhaus [**3**, (20.12)] $Cs\mathfrak{P}_1$ contains a normal subgroup $\mathfrak{S}$ such that $Cs\mathfrak{P}_1 = \mathfrak{Z}^{\#}\mathfrak{S}$ with $\mathfrak{S} \cap \mathfrak{Z}^{\#} = \mathfrak{E}$. Now $\mathfrak{Z}^{\#}$ contains $\mathfrak{P}_1$. If $\mathfrak{Z}^{\#} = \mathfrak{P}$, then $Cs\mathfrak{P}_1$ is p-nilpotent. Then $B^{-1}A^{-1}BA$ belongs to $\mathfrak{S}$ whose order is prime to p. Consider the action of $B^{-1}A^{-1}BA$ on $\alpha(P_1)$. B on $\alpha(P_1)$ is a p-cycle and A on $\alpha(P_1)$ leaves only the point 1 fixed. Since $Cs\mathfrak{P}_1/\mathfrak{N} \cap Cs\mathfrak{P}_1$ is solvable, $B^{-1}A^{-1}BA$ is a p-cycle on $\alpha(P_1)$. This is a contradiction. Therefore we must have that $\mathfrak{Z}^{\#} = \mathfrak{P}_1$. So $Cs\mathfrak{P}_1$ is a direct product of $\mathfrak{P}_1$ and $\mathfrak{S}$. Let $\mathfrak{P}_0 = \langle B' \rangle$ and $\mathfrak{T}$ be a Sylow p-subgroup and a Sylow p-complement of $\mathfrak{S}$ respectively. Put $\mathfrak{U} = \mathfrak{S} \cap \mathfrak{N}$. We may assume that $\mathfrak{T}$ contains A as a central element. Then $\mathfrak{U}$ is contained in CsA. If $\mathfrak{P}_0$ is not contained in $Cs\mathfrak{U}$, then $B'^{-1}A^{-1}B'A$ gives the same contradiction as above. So $\mathfrak{P}_0$ is contained in $Cs\mathfrak{U}$. Hence $\mathfrak{P}_0$ is normal in $\mathfrak{S}$. Therefore $Ns\mathfrak{P}$ contains A. Since $\mathfrak{G} = Ns\mathfrak{P}CsA$, this shows that core $(Ns\mathfrak{P})$ contains A. This contradiction shows that $\mathfrak{P}_1 = \mathfrak{E}$. Thus p divides $|\mathfrak{G}|$ only to the first power.

(c) Since $\mathfrak{G} = CsA\langle BC \rangle = CsA \cdot CsBC$, we have that $\chi(A)\chi(BC) = \chi(E)\chi(ABC)$ for every irreducible character χ of $\mathfrak{G}$. Since $\mathfrak{G} : CsA = pq$, if $\chi \neq 1_\mathfrak{G}$, the principal character of $\mathfrak{G}$, and if $(\chi(E), pq) = 1$, then by a theorem of Burnside [**3**, (18.1)] we may assume that $\chi(A) = 0$. This implies that $\chi(ABC) = 0$. If $\chi(E)$ is divisible by p, then by a theorem of Brauer [1], [3, (17.4)] we have that $\chi(BC) = 0$. This implies that $\chi(ABC) = 0$. Now we have that

$$\sum_\chi \chi(E)\chi(ABC) = 0. \tag{3}$$

So $-1 = \sum_{\chi \neq 1_\mathfrak{G}} \chi(E)\chi(ABC)$. Now if $\chi(ABC) \neq 0$ for $\chi \neq 1_\mathfrak{G}$, then $(\chi(E), p) = 1$ and $(\chi(E), pq) \neq 1$. Thus we see that $\chi(E)$ is divisible by q. So we get the contradiction $1 \equiv 0 \pmod{q}$.

PROOF OF PROPOSITION 6. (a) Assume that $\mathfrak{G}$ is a minimal counterexample to Proposition 6. By (a) of the proof of Proposition 5 we may assume that $\mathfrak{G}$ is a transitive permutation group on $\Omega = \{1, \ldots, pq\}$, that CsA is the stabilizer of the point 1 in Ω, and that $\langle BC \rangle$ is a regular subgroup of $\mathfrak{G}$ with $BC = CB$, $|\langle B \rangle| = p$ and $|\langle C \rangle| = q$. If CsA is not maximal in $\mathfrak{G}$, then let $\mathfrak{M}$ be a maximal subgroup of $\mathfrak{G}$ containing CsA. We may assume that $\mathfrak{M} = CsA\langle B \rangle$. Put $\mathfrak{N}$ = core $\mathfrak{M}$. By (iv) of §2 $\mathfrak{G}/\mathfrak{N}$ is solvable. Hence $\mathfrak{N}$ is not solvable. If $\mathfrak{N}$ contains A, then consider $\mathfrak{N}\langle C \rangle = (\mathfrak{N} \cap CsA)\langle BC \rangle$. If $\mathfrak{G} \neq \mathfrak{N}\langle C \rangle$, then by the minimality of $\mathfrak{G}$ $\mathfrak{N} \cdot \langle C \rangle$ is solvable. This is a contradiction. Thus $\mathfrak{G} = \mathfrak{N} \cdot \langle C \rangle$. This implies that $\mathfrak{N} = \mathfrak{M}$. Let $\mathfrak{X}$ be the core of CsA in $\mathfrak{M}$. Then by (iv) of §2 $\mathfrak{M}/\mathfrak{X}$ is solvable. Thus core (CsA) contains $\mathfrak{M}^*$. This is a contradiction, so A does not belong to $\mathfrak{N}$. If $\mathfrak{G} = Cs(A \div \mathfrak{N})$, where $Cs(A \div \mathfrak{N})/\mathfrak{N} = Cs(A\mathfrak{N})$, then $\langle A \rangle \mathfrak{N}$ is normal in $\mathfrak{G}$. Since $\mathfrak{N}$ is the core of $\mathfrak{M}$, this is a contradiction. If $\mathfrak{G} \neq Cs(A \div \mathfrak{N})$, then

$$Cs(A \div \mathfrak{N}) = CsA(Cs(A \div \mathfrak{N}) \cap \langle BC \rangle) = CsA \cdot \langle B \rangle.$$

Let $\mathfrak{Y}$ be the core of CsA in $Cs(A \div \mathfrak{N})$. Then by (iv) of §2 $Cs(A \div \mathfrak{N})/\mathfrak{Y}$ is solvable. Thus CsA contains $\mathfrak{N}^*$. This is a contradiction. Therefore CsA is maximal in $\mathfrak{G}$. So by a theorem of Schur [12, (25.3)] $\mathfrak{G}$ is doubly transitive on Ω. We may assume that $|\langle A \rangle| = r$ is a prime. Obviously we have that $\alpha(A) = 1$. So, in particular, $pq \equiv 1 \pmod{r}$. Assume that $p > q$. Then by (b) of the proof of Proposition 5 we obtain that p divides $|\mathfrak{G}|$ only to the first power.

(b) By Proposition 5 $\mathfrak{G}$ is not simple. Since $\mathfrak{G}$ is doubly transitive on Ω, every nontrivial normal subgroup $\mathfrak{N}$ of $\mathfrak{G}$ is transitive on Ω. Therefore $\mathfrak{N}$ is not solvable and contains B. Furthermore $\mathfrak{G}$ contains a unique minimal normal subgroup $\mathfrak{L}$. $\mathfrak{L}$ is simple.

(c) We have that $CsB = \mathfrak{X}\langle BC \rangle$, where $\mathfrak{X}$ is a subgroup of CsA. $Ns\mathfrak{X}$ contains A and B. Suppose that $\mathfrak{X} \neq \mathfrak{E}$. If $Ns\mathfrak{X}$ contains C, then $\mathfrak{X}$ is contained in core (CsA). This is a contradiction. Therefore $Ns\mathfrak{X} \cap CsB = X\langle B \rangle$. Since CsB is transitive on Ω, by a theorem of Jordan [12, (3,5)] we get that $|\alpha(X)| = p$. Since $Ns\mathfrak{X}$ contains A, by (iv) in §2 $Ns\mathfrak{X}/\mathfrak{Y}$ is solvable where $\mathfrak{Y}$ is the kernel of the permutation representation of $Ns\mathfrak{X}$ on $\alpha(X)$. Since $Cs\mathfrak{Y} \cap Ns\mathfrak{X}$ contains A, it also contains B. Thus we have that $A^{-1}BA = A^a Y_1$, where Y_1 is an element of $\mathfrak{Y}$ and a is an integer. Obviously this implies that $Y_1 = E$. So $Ns\langle B \rangle$ contains A, and hence core $(Ns\langle B \rangle)$ contains A, and hence core $(Ns\langle B \rangle) \neq \mathfrak{E}$. So NsB contains $\mathfrak{L}$. Since $\mathfrak{L}$ is simple, is nonabelian and contains B, this is a contradiction. So $\mathfrak{X} = \mathfrak{E}$ and $CsB = \langle BC \rangle$. Now $\mathfrak{L}CsB = \mathfrak{L}\langle C \rangle$ is normal in $\mathfrak{G} = \mathfrak{L}Ns\langle B \rangle$. If $\mathfrak{G} \neq \mathfrak{L}\langle C \rangle \langle A \rangle$, then by the minimality of $\mathfrak{G}$ $\mathfrak{L}\langle C \rangle \langle A \rangle$ is solvable. This is a contradiction. Hence we obtain that $\mathfrak{G} = \mathfrak{L}\langle C \rangle \langle A \rangle$. If $\mathfrak{L}$ contains A or C, then $\mathfrak{L}$ is a unique nontrivial normal subgroup of $\mathfrak{G}$. If $\mathfrak{L}$ contains neither A nor C, then $\mathfrak{G}/\mathfrak{L}$ has order qr and $\mathfrak{L}\langle C \rangle / \mathfrak{L}$ is normal in $\mathfrak{G}/\mathfrak{L}$. Therefore, in any case, if $\chi(A)$ is a nonzero multiple of $\chi(E)$ for an irreducible character χ of $\mathfrak{G}$, then we have that $\chi(E) = 1$.

(d) Put $Ns\langle B \rangle = \langle M \rangle CsB$, where $\langle M \rangle$ is a subgroup of CsA. Put $|M| = m$ and $p - 1 = mt$. Then we partition $\langle C \rangle$ into $\langle M \rangle$-conjugacy classes: $\mathfrak{C}_1 = \{E\}$, $\mathfrak{C}_2, \ldots, \mathfrak{C}_l$. Obviously all the classes $\mathfrak{C}_i$ $(i > 1)$ consist of the same number, say τ, of elements. τ is a common divisor of $q - 1$ and m; $\tau_1 = q - 1$. Put $\tau_1 = 1$, $\tau_i = \tau$ $(i > 1)$ and $t_i = \tau_i t$ $(i \geq 1)$. Then by Brauer [2] we have the following situation. $\mathfrak{G}$ has l blocks $\mathscr{B}_1, \ldots, \mathscr{B}_l$ of irreducible characters of $\mathfrak{G}$ of full p-defect. Each block $\mathscr{B}_i$ consists of $p - 1/t_i$ nonexceptional characters ζ and t_i exceptional characters ξ_j $(1 \leq j \leq t_i)$. The ζ take the same value for each element of $\mathfrak{G}$ whose order is prime to p. Furthermore, to each block $\mathscr{B}_i$, there corresponds an irreducible character θ_i of $\langle C \rangle$ which satisfies the following condition:

(4) $$\zeta(BC^k) = \delta \sum_{j=0}^{\tau_i - 1} \theta_i(M^{-j}C^k M^j),$$

where $\{M^{-j}C^k M^j, 0 \leq j \leq \tau_{i-1}\}$ is the $Ns\langle B \rangle$-conjugacy class containing C^k $(0 \leq k \leq q - 1)$ and $\delta = \pm 1$.

(e) Assume that $i > 1$ and $\tau_i > 1$. Then add (4) for all the representatives of $Ns\langle B \rangle$-conjugacy classes $\neq \{E\}$ of $\langle C \rangle$. Then we obtain that

(5) $$\sum \zeta(BC^k) = -\delta.$$

Now taking BC instead of B, observe the equation (2) in §1:

$$\chi(A)\chi(BC) = \chi(E)\chi(ABC).$$

Let $\chi = X_0$, the doubly transitive character of $\mathfrak{G}$ on Ω. Then we have that $X_0(ABC) = 0$. So the order of ABC is prime to p. Now if χ has p-defect 0, then by a theorem of Brauer [1] we have that $\chi(BC) = 0$ and hence $\chi(ABC) = 0$. If χ belongs to $\mathscr{B}_i$ and if $\chi = \zeta$, then by (4) or (5) $\chi(A)$ is divisible by $\chi(E)$. Hence if χ is nonlinear, by a theorem of Burnside [3, (18.1)] and (c) we have that $\chi(A) = 0$ and hence $\chi(ABC) = 0$. Suppose that χ belongs to $\mathfrak{B}_i$ and that $\chi = \xi_j$. We have that

$$\xi_j(A)\xi_j(BC^k) = \xi_j(E)\xi_j(ABC^k).$$

So if $\xi_j(A) \neq 0$, then we have that $\xi_1(BC^k) = \xi_2(BC^k) = \ldots$. This is a contradiction. So we have that $\xi_j(A) = 0$ and hence $\xi_j(ABC) = 0$. Thus we obtain that

$$\sum \chi(ABC)\overline{\chi(ABC)} = \sum_{\chi(E) = 1} \chi(ABC)\overline{\chi(ABC)} = |Cs(ABC)|.$$

Since $\alpha(ABC) = 1$, $|Cs(ABC)|$ is divisible by r. If $|Cs(ABC)| = r$, then $|CsA| = r$ and hence $|\mathfrak{G}| = pqr$. This is a contradiction. If $|ABC| = q$ or qr, then there exists an element X of order q of CsA such that $|Cs(AX)| = qr$. This implies that $\langle X \rangle$ is a Sylow q-subgroup of CsA. Furthermore it follows that $\mathfrak{G}/\mathfrak{L}$ is cyclic of order qr and that $AC\mathfrak{L}$ is a generator of $\mathfrak{G}/\mathfrak{L}$. Since $\mathfrak{G} = CsA \cdot \mathfrak{L}$, $CsA = \langle AX \rangle \cdot \mathfrak{L} \cap CsA$. Hence we have that $Ns\langle X \rangle \cap CsA = CsX \cap CsA$. Now we may assume that $XC = CX$. Then put $CsC = \mathfrak{X} \langle BC \rangle$, where $\mathfrak{X}$ is a subgroup of CsA. X belongs to $\mathfrak{X}$. Since CsC is transitive on Ω, by a theorem of Jordan [**12**, (3.5)] $Ns\mathfrak{X} \cap CsC = \mathfrak{X} \langle C \rangle$ is transitive on $\alpha(\mathfrak{X})$. Thus $|\alpha(\mathfrak{X})| = q$. Since $\mathfrak{X}$ is a subgroup of CsA, it follows that $q \equiv 1 \pmod{r}$. Now since CsA is transitive on $\Omega - \{1\}$, by a theorem of Jordan [**12**, (3.5)] $CsA \cap CsX$ is transitive on $\alpha(X) - \{1\}$. Thus $|\alpha(X) - \{1\}| = r \geqq q - 1$. So we obtain that $q = 3$ and $r = 2$.

By a theorem of Feit-Thompson [**4**] and by a theorem of Glauberman [**6**], A belongs to the center of $\mathfrak{G}$. This is a contradiction.

Bibliography

1. R. Brauer and C. Nesbitt, *On the modular characters of groups*, Ann. of Math. (2) **42** (1941), 556–590. MR **2**, 309.

2. R. Brauer, *On groups whose order contains a prime number to the first power*, I, Amer. J. Math. **64** (1942), 401–420. MR **4**, 1.

3. W. Feit, *Characters of finite groups*, Benjamin, New York, 1967. MR **36** #2715.

4. W. Feit and J. G. Thompson, *Solvability of groups of odd order*, Pacific J. Math. **13** (1963), 775–1029. MR **29** #3538.

5. B. Fischer, *ᚠ-Gruppen endlicher Ordnung*, Arch. Math. **16** (1965), 330–336. MR **33** #1364.

6. G. Glauberman, *Central elements in core-free groups*, J. Algebra **4** (1966), 403–420. MR **34** #2681.

7. B. Huppert, *Endliche Gruppen*. I, Die Grundlehren der math. Wissenschaften, Band 134, Springer-Verlag, Berlin, 1967. MR **37** #302.

8. N. Ito, *On transitive permutation groups of prime degree*, Sûgaku **15** (1963/64), 129–141. (Japanese) MR **29** #3531.

9. ———, *On finite groups with given conjugate types*. III Math. Z.**117** (1970), 267–271.

10. J. Szép. *Zur Theorie der faktorisierbaren Gruppen*, Acta Sci. Math. (Szeged) **16** (1955), 54–57. MR **17**, 455.

11. ———, *Sui gruppi fattorizzabili*, Rend. Sem. Mat. Fis. Milano **38** (1968), 228–230. MR **39** #1542.

12. H. Wielandt, *Finite permutation groups*, Lectures, University of Tübingen, 1954/55; English transl., Academic Press, New York, 1964. MR **32** #1252.

University of Illinois at Chicago Circle

Lattices Over Orders

H. Jacobinski

Let o be a Dedekind ring, k its quotient field and A/k a separable finite dimensional k-algebra. An o-order in A is a subring R which at the same time is a finitely generated o-module and such that moreover $1 \in R$ and $kR = A$. An important example is the integral group ring oG, G a finite group, which is an o-order in kG where char $k \nmid |G|$.

An integral representation of an order R is afforded by an R-lattice M, i.e. a finitely generated left R-module which is torsion-free as an o-module. In particular, integral representations of a finite group G correspond to oG-lattices.

It is well known that R-lattices have a much more complicated behaviour than A-modules. This is mainly due to the fact that they do not have unique decomposition, i.e. the analogon of the Krull-Schmidt theorem does not hold for R-lattices. Even the weaker cancellation property fails, that is $M \oplus X \cong N \oplus X$ need not imply $M \cong N$.

If, however, we replace o by its completion o_p at a prime p and R correspondingly by $R_p = o_p \otimes R$, then the category of R_p-lattices has unique decomposition (Reiner [**2**, p. 171]). This fact motivates the introduction of genera of R-lattices, which is due to Maranda. Two R-lattices M,N belong to the same genus Γ, notation $M \sim N$, if $M_p \cong N_p$ for all primes p. In fact it is sufficient that this isomorphism holds for a finite set of primes U, depending on R.

In order to translate properties of R_p-lattices to those of R-lattices, one needs a description of the isomorphism classes in a genus. As an introduction, let us first consider the well-known case of lattices over a Dedekind ring, i.e. $R = o$. Since o_p-lattices are free, two o-lattices M and N are in the same genus, if the vector spaces kM and kN are isomorphic. Replacing N by an isomorphic lattice, we can thus

---

AMS 1970 *subject classifications.* Primary 16A18, 16A26, 20C10; Secondary 16A50.

assume $kM = kN$. Then M and N are isomorphic if and only if the order ideal [2, p. 162] from M to N is principal. Consequently, the isomorphism classes in the genus $\Gamma(M)$ are in 1-1 correspondence with the (absolute) ideal classes in o.

It turns out that this description of isomorphism classes in a genus can be generalized to arbitrary orders R, provided k is an algebraic number field (Jacobinski [1, §2]). The proofs, however, are much more difficult and use rather deep results from class-field theory. The order-ideal has to be replaced by the norm-ideal, to be defined later, and one has to use finer ideal classes of the type occurring in classical class-field theory instead of the absolute ideal classes.

Let U be a finite set of primes p, containing all primes p such that R_p is not a maximal order. We define the clone $\Lambda(M)$ of M to be the set of all R-lattices X, such that $kX = kM$ and $X_p = M_p$ for all $p \in U$. Then $\Lambda(M) \subset \Gamma(M)$ and it is easily seen that every lattice in $\Gamma(M)$ is isomorphic to a lattice in $\Lambda(M)$.

Now we consider the endomorphism ring $\mathrm{End}_A(kM)$; let C be the integral closure over o of its center. This is a direct sum of Dedekind rings. Let I_M be the group of all invertible fractional C-ideals, that are prime to all p in U.

For two lattices X, Y in the clone $\Lambda(M)$ we define the norm-ideal $n(X, Y)$ to be the C-ideal generated by the reduced norms of all monomorphisms in $\mathrm{Hom}_R(X, Y)$. Since $X_p = Y_p$ for all p in U, $n(X, Y)$ is prime to all p in U. Thus, if we fix X, the assignment $(X, Y) \to n(X, Y)$ induces a map

$$\varphi_X \colon \Lambda(M) \to I_M,$$

which is surjective, but not injective. The important property of φ_X is that isomorphism classes in $\Lambda(M)$ are characterized by their image under φ_X. One way this is easy to see. For suppose that X and Y are isomorphic, $Y = X\alpha$ with $\alpha \in \mathrm{End}_A(kX)$. Because of the multiplicativity of the reduced norm, this implies that $\varphi_X(Y) = (n(\alpha))$. Thus if H_X is the subgroup of I_M generated by all ideals $(n(\alpha))$ with α a monomorphism in $\mathrm{Hom}_R(X, X)$, we see that $Y \cong X$ implies $\varphi_X(Y) \subset H_X$. The difficult part is to show the converse of this and that H_X is the same for all $X \in \Lambda(M)$. Here we need the assumption that k is an algebraic number field and results from class-field theory. Moreover M has to satisfy the Eichler condition (i.e. none of the simple algebras in $\mathrm{End}_A(kM)$ is a totally definite quaternion skew-field).

THEOREM [1, p. 7]. *Let M be an R-lattice satisfying the Eichler condition and let H_M be the subgroup of I_M generated by principal ideals $(n(\alpha))$, where α is a monomorphism $M \to M$. Then two lattices in $\Lambda(M)$ are isomorphic if and only if $n(X, Y) \in H_M$ and the isomorphism classes in $\Gamma(M)$ are in 1-1 correspondence with the elements of the factor group $V_M = I_M/H_M$.*

Let us first compare this with the classical results on o-lattices. In this case we can take $U = \emptyset$ and the norm ideal reduces to the order ideal. H_M is the group of all principal ideals and V_M the absolute class group of o. Lattices over a Dedekind ring have the special property that every genus contains a free module F. The ideal class of $n(F, M)$ is the Steinitz invariant of M and completely describes the isomorphism class of M in its genus. In the case of an arbitrary order R, a genus need not contain a free or otherwise distinguished module, so that it does not seem possible to generalize the Steinitz invariant.

Returning to the general case, we mention that H_M is of the same type as the groups of ideals encountered in class field theory. One can show that H_M contains the ray modulo a certain conductor which depends only on R and not on M.

The theorem above has a number of interesting applications to the global behaviour of R-lattices. We are now going to describe the two most important ones.

We call an R-lattice X a local direct factor of M if X_p is isomorphic to a direct factor of M_p for all p (in fact it is sufficient that this holds for all p in U). The question then arises, under what supplementary conditions is X then isomorphic to a direct factor of M? One sufficient condition can be obtained by specialization from a more general theorem of Serre. The following theorem however is much stronger.

THEOREM [1, p. 14]. *Let X be a local direct factor of M and suppose that every irreducible A-module that occurs in kX occurs with greater multiplicity in kM. Then X is isomorphic to direct factor of M.*

As a consequence of this theorem for instance, let P be a projective R-lattice which is locally free everywhere, of rank $s + 1$, say. Then a free R-module F_s of rank s is a local direct factor of P and the theorem yields $P \cong F_s \oplus I$, where I is a projective ideal. In the case $R = oG$ this was shown by Swan (see [2, p. 187]).

As a second application of the theorem on genera, we consider cancellation, i.e. whether $M \oplus X \cong N \oplus X$ implies $M \cong N$. To begin with, we mention an anti-cancellation theorem. Let O be a maximal order containing R. Two R-lattices M,N belong to the same restricted genus [1, p. 10] if $M \sim N$ and moreover $OM \cong ON$. We denote this by $M \approx N$. Then one can show [1, p. 11] that there exists a lattice T, depending only on R such that

$$M \oplus T \cong N \oplus T \Leftrightarrow M \approx N,$$

provided M satisfies the Eichler condition. This shows that cancellation is permitted for arbitrary R-lattices M,X, satisfying the Eichler condition, if and only if every restricted genus contains only one isomorphism class. This is true for special orders, e.g. maximal orders but not in general.

Thus in order for cancellation to be permitted, the couple M,X has to satisfy some supplementary condition. This condition can be formulated in terms of the groups H_M and $H_{M \oplus X}$. Clearly $M \oplus X \cong N \oplus X$ implies $M \sim N$, and we can suppose $N \in \Lambda(M)$ and $N \oplus X \in \Lambda(M \oplus X)$. Then $n(M \oplus X, N \oplus X) = n(M,N)$ and the theorem on isomorphism classes in a genus yields that cancellation is possible if and only if $H_M = H_{M \oplus X}$. Using this condition, we can show [1, p. 15]

THEOREM. *Suppose that M satisfies the Eichler condition and that X is a local direct factor of a finite direct sum $M \oplus M + \cdots \oplus M$. Then $M \oplus X \cong N \oplus X$ implies $M \cong N$.*

We mention an interesting application of this theorem, concerning cancellation in the category of finitely generated projective oG-modules, where G is a finite group. This problem was first studied by Swan, who showed by an example ($o = Z$, G a generalized quaternion group) that cancellation is not always permitted. Using the above theorem one can show [1, p. 19]

THEOREM. *Suppose that no rational prime dividing the order of G is a unit in o and that the group ring oG satisfies the Eichler condition. Then cancellation is always permitted in the category of finitely generated projective oG-modules.*

REFERENCES

1. H. Jacobinski, *Genera and decompositions of lattices over orders*, Acta Math. **121** (1968), 1–29. MR **40** #4294.

2. I. Reiner, *A survey of integral representation theory*, Bull. Amer. Math. Soc. **76** (1970), 159–227.

CHALMERS UNIVERSITY OF TECHNOLOGY, GÖTEBORG

Faithful Representations of p Groups at Characteristic p

G. J. Janusz

Our object is to investigate representations of a p group G over fields K of characteristic p. K is assumed to be infinite. We shall consider questions of the general form, "If G has only a finite number of faithful indecomposable representations of a special kind, then what can be said about G?"

The earliest result of this sort was given by Higman [4] who proved that G must be cyclic if there are only finitely many inequivalent indecomposable K-representations of G.

If the group G is the noncyclic group of order four, the representations are known by the results of Bašev [1], Conlon [2], Heller-Reiner [3] and also Johnson [6]. In particular, it turns out in this case that G has only two inequivalent indecomposable representations in odd dimensions ≥ 3.

Our first result shows that no other abelian groups can have this kind of property.

THEOREM 1. *Let G be an abelian p group which is neither cyclic nor of order four. Let K be an infinite field of characteristic p. For any integer $d \geq 2$, G has infinitely many inequivalent indecomposable K representations in dimension d. If $p(d-1) \geq$ exponent of G then infinitely many of these represent G faithfully.*

Next we consider nonabelian groups. If a $K(G)$-module, M, affords a faithful representation of G but no proper submodule and no proper homomorphic image of M affords a faithful representation of G, we shall call M a *fundamental module*.

By results in [5] it follows that a fundamental module is isomorphic to a principal left ideal in $K(G)$. In particular the dimension is bounded by $|G|$. One may think of fundamental modules as small, yet the next theorem shows they are plentiful.

AMS 1969 *subject classifications.* Primary 2080, 2040.

THEOREM 2. *Let G be a nonabelian p group and K an infinite field of characteristic p. Then there exist infinitely many nonisomorphic fundamental modules.*

Finally, we consider what groups G can have only a finite number of faithful indecomposable K representations in one particular dimension, d. If G is abelian, then by Theorem 1 G is cyclic or has order four. However, if G is nonabelian there do exist nontrivial examples. Namely, let G be either the dihedral group or generalized quaternion group of order 2^{n+2}. Then G has precisely two faithful representations in dimension $1 + 2^n$ over any field K of characteristic 2 provided $GF(4) \subseteq K$ in the case of the quaternion group of order eight.

It seems to me that groups with this property are very scarce. To support this feeling we prove that under some additional hypothesis only a finite number of groups can have this property once p and d are given. More precisely we have this last result.

THEOREM 3. *Let G be a nonabelian p group and K an **infinite** field of characteristic p. Suppose $K(G)$ has only a finite number ($\neq 0$) of fundamental modules of dimension d. Then* (i) *the number of generators of $G \leq \frac{1}{2}(d + 1)(d - 2)$ and* (ii) $|G| \leq f_p(d)$ *for some function f_p which depends only on the prime p.*

REFERENCES

1. V. A. Bašev, *Representations of the group $Z_2 \times Z_2$ in a field of characteristic* 2, Dokl. Akad. Nauk SSSR **141** (1961), 1015–1018 = Soviet Math. Dokl. **2** (1961), 1589–1592. MR **24** #A1944.

2. S. B. Conlon, *Certain representation algebras*, J. Austral. Math. Soc. **5** (1965), 83–99. MR **32** #2494.

3. A. Heller and I. Reiner, *Indecomposable representations*, Illinois J. Math. **5** (1961), 314–323. MR **23** #A222.

4. D. G. Higman, *Indecomposable representations at characteristic p*, Duke Math. J. **21** (1954), 377–381. MR **16**, 794.

5. G. J. Janusz, *Faithful representations of p groups at characteristic p*, 1, J. Algebra **15** (1970), 335–351.

6. D. L. Johnson, *Odd dimensional representations of $Z_2 \times Z_2$*, J. London Math. Soc. **15** (1969).

UNIVERSITY OF ILLINOIS, URBANA

The Reflection Character of a Finite Group with a (B,N) Pair

Robert Kilmoyer

1. Let (G,B,N,R) be a finite group with a (B,N) pair [1], [6]. Thus the Weyl group $W = B/B \cap N$ of G has the presentation with generators R and relations

$$r^2 = 1, \quad r \in R,$$

$$(rsr\cdots)_{n_{r,s}} = (srs\cdots)_{n_{r,s}}, \quad r,s \in R,$$

where $n_{r,s} = |\langle rs \rangle|$ and the notation $(aba\cdots)_n$ means an alternating product of a's and b's until n factors are obtained. We assume throughout that the Coxeter system (W,R) is irreducible. This means that it is impossible to partition R into two non-empty disjoint subsets R_1 and R_2 such that $[R_1, R_2] = 1$.

Let k be a field of characteristic zero and kG the group algebra of G over k. Then $e = |B|^{-1} \sum_{x \in B} x$ is an idempotent in kG and kGe affords the induced representation 1_B^G where 1_B denotes the trivial representation of B. $H_k(G,B) = ekGe$ is called the *Hecke algebra* of G relative to B over k. Thus $H_k(G,B)$ is naturally isomorphic with the centralizer algebra $\mathrm{Hom}_{kG}(kGe, kGe)$. Put $a_w = |B|^{-1} \sum_{x \in BwB} x$, $w \in W$. Since the (B,B) double cosets of G may be indexed by W it follows that $\{a_w | w \in W\}$ is a k-basis of $H_k(G,B)$. One knows [3], [5] that $H_k(G,B)$ has the presentation with generators $\{a_r | r \in R\}$ and defining relations:

$$a_r^2 = q_r \cdot 1 + (q_r - 1)a_r, \quad r \in R,$$

$$(a_r a_s a_r \cdots)_{n_{r,s}} = (a_s a_r a_s \cdots)_{n_{r,s}}, \quad r,s \in R,$$

where for any $w \in W$ we have put $q_w = [B : B \cap wBw^{-1}]$. The elements q_r, $r \in R$ are called the *index parameters* of G.

If k is algebraically closed, one knows [2] that each irreducible character φ of

---
AMS 1970 subject classifications. Primary 20C15.

$H_k(G,B)$ is the restriction to $H_k(G,B)$ of a unique irreducible character ζ of G such that $(\zeta, 1_B^G) \neq 0$. Moreover

$$\zeta(1) = \frac{\varphi(e)[G:B]}{\sum_{w \in W} q_w^{-1} \varphi(a_w^{-1}) \varphi(a_w)}.$$

Also in the case when k is algebraically closed one knows [3] that $kW \cong H_k(G,B)$.

In order to compare kW with $H_k(G,B)$ Tits introduced the generic ring A of (W,R) defined as follows: Let $\{u_r | r \in R\}$ be indeterminates such that $u_r = u_s$ if and only if r and s are conjugate in W. Let $O = k[u_r; r \in R]$ be the polynomial ring, and let A be the O-algebra with identity having the presentation with generators $\{a_r; r \in R\}$ and defining relations:

(1)
$$a_r^2 = u_r \cdot 1 + (u_r - 1)a_r, \quad r \in R,$$
$$(a_r a_s a_r \cdots)_{n_{r,s}} = (a_s a_r a_s \cdots)_{n_{r,s}}, \quad r, s \in R.$$

If $w = r_1 r_2 \cdots r_m$ is a reduced expression for an element $w \in W$, put $a_w = a_{r_1} a_{r_2} \cdots a_{r_m}$. Then a_w is independent of the choice of the reduced expression, and the set $\{a_w; w \in W\}$ forms a free O-basis for A as an O-module. It is clear that the specialization $u_r \to q_r$ yields $H_k(G,B)$, while the specialization $u_r \to 1$ yields kW.

Now the Weyl group W of G has a distinguished irreducible representation as a group generated by reflections on a Euclidean vector space of dimension $|R|$. Moreover, it is known that the exterior powers of this representation are irreducible and distinct. We call this distinguished representation of W the *reflection representation*; and call the exterior powers of it the *compounds* of the reflection representation. The object of this note is to show how the analogues of the reflection representation and its compounds can be constructed for the generic ring A and to give some results on the corresponding irreducible characters of the group G.

2. We shall use the following notation: Q = the field of rational numbers, $O = Q[u_r; r \in R]$, K = the quotient field of O, $\bar{K}$ = an algebraic closure of K, and O^* = the integral closure of O in $\bar{K}$. Let V be a $\bar{K}$-vector space of dimension $l = |R|$ and let $\{c_{r,s}; r, s \in R\}$ be any elements of $\bar{K}$ such that

$$c_{r,r} = u_r + 1, \quad r \in R,$$
$$c_{r,s} = c_{s,r} = 0, \quad r, s \in R \quad \text{if } n_{r,s} = 2,$$
$$c_{rs} c_{sr} = u_r + u_s + 2\sqrt{u_r u_s} \cos \frac{2\pi}{n_{r,s}} \quad \text{if } n_{r,s} > 2.$$

LEMMA. *There exists a nonzero, symmetric bilinear form B on V, unique up to a scalar multiple such that*

$$c_{r,s} = \frac{u_{r+1} B(\alpha_r, \alpha_s)}{B(\alpha_r, \alpha_r)}, \quad r, s \in R.$$

Now define $\chi_r \in \operatorname{End}_{\bar{K}} V$ by

$$\chi_r(\xi) = u_r \xi - \frac{(u_r + 1)B(\alpha_r,\xi)}{B(\alpha_r,\alpha_r)} \alpha_r, \quad r \in R, \xi \in V,$$

and

$$\chi_r^{(k)} \in \operatorname{End}_{\bar{K}} \Lambda^k V \text{ by } \chi_r^{(k)} = u_r^{-(k-1)} \Lambda^k \chi_r, \quad 0 \leq k \leq l.$$

It can be shown then that the defining relations (1) are satisfied with $\chi_r^{(k)}$ in place of a_r, $r \in R$. Thus there exists a unique representation $\pi^{(k)}: A^{\bar{K}} \to \operatorname{End}_{\bar{K}} \Lambda^k V$ ($0 \leq k \leq l$) such that $\pi^{(k)}(a_r) = \chi_r^{(k)}$, $r \in R$. We call the $\pi^{(k)}$ the *compounds* of the *reflection representation* $\pi = \pi^{(1)}$.

3. The detailed proofs of the results in this section will appear elsewhere.

THEOREM 1. (i) *The bilinear form B is nondegenerate.*

(ii) *The representations $\pi^{(k)}$ are irreducible and pairwise inequivalent* ($0 \leq k \leq l$).

(iii) *The $\pi^{(k)}$ are all defined over the ring*:
  $O[(u_r u_s)^{1/2}]$, *if W is dihedral of order* 12;
  $O[(2u_r u_s)^{1/2}]$, *if W is dihedral of order* 16;
  O *in all other cases.*

The statement "$\pi^{(k)}$ is defined over the ring O'" means that a basis for $\Lambda^k V$ can be chosen so that the matrices of the elements $\pi^{(k)}(a_w)$, $w \in W$ relative to this basis have all their coefficients in the ring O'.

THEOREM 2. *Let (G,B,N,R) be a finite group with a (B,N) pair having the Coxeter system (W,R). Let $\zeta^{(k)}$ be the irreducible character of G corresponding to $\pi^{(k)}$ ($0 \leq k \leq l$), and let P be a parabolic subgroup of G, then*

(i) $(\zeta^{(k)}, 1_P^G) = \binom{l-t}{k}$, $l = \operatorname{rank} G$, $t = \operatorname{rank} P$;

(ii) *If W is not dihedral of order* 12 *or* 16, *then the reflection character $\zeta = \zeta^{(1)}$ of G is uniquely determined by the multiplicities of* (i);

(iii) *The characters $\zeta^{(k)}$ are all afforded by rational representations of G.*

We have determined the degree $\zeta(1)$ for each finite (B,N) pair as a function of the index parameters using a case by case analysis. It is interesting that if $G = G(q)$ is a Chevalley group whose Dynkin diagram is simply laced (i.e. of type A_l, D_l, or E_l), then $\zeta(1) = \sum_{i=1}^{l} q^{m_i}$ where $\{m_1, m_2, \ldots, m_l\}$ are the exponents of (W,R).

REFERENCES

1. N. Bourbaki, *Groupes et algèbres de Lie*. Chaps. 4, 5, 6, Actualités Sci. Indust., no. 1337, Hermann, Paris, 1968. MR **39** #1590.

2. C. W. Curtis and T. V. Fossum, *On centralizer rings and characters of representations of finite groups*, Math. Z. **107** (1968), 402–406. MR **38** #5946.

3. N. Iwahori, *Generalized Tits systems on p-adic semi-simple groups*, Proc. Sympos. Pure Math., vol. 9, Amer. Math. Soc., Providence, R.I., 1966, pp. 71–83. MR **35** #6693.

4. R. Kilmoyer, *Some irreducible complex representations of a finite group with a BN-pair*, Ph.D. Dissertation, M.I.T., Cambridge, Mass., 1969.

5. H. Matsumoto, *Générateurs et relations des groupes de Weyl généralisés*, C. R. Acad. Sci. Paris **258** (1964), 3419–3422. MR **32** #1294.

6. J. Tits, *Théorème de Bruhat et sous-groupes paraboliques*, C. R. Acad. Sci. Paris **254** (1962), 2910–2912. MR **25** #2149.

CLARK UNIVERSITY

A Characterization of the Alternating Groups

Takeshi Kondo

In [3], I proved the following Theorem A. Let A_m be the alternating group on m letters $\{1,2,\ldots,m\}$ ($m \geq 4$). Put $m = 4n + r$ where n is a positive integer ≥ 1 and $0 \leq r \leq 3$. Let $\tilde{\alpha}_n$ be an involution of A_m as follows:

$$\tilde{\alpha}_n = (1,2)(3,4)\cdots(4n-1,4n).$$

For $r = 1, 2$ and 3, we denote by $\tilde{H}(n,r)$ the centralizer of $\tilde{\alpha}_n$ in A_m.

THEOREM A. *Let $G(n,r)$ ($n \geq 1$ and $r = 1, 2$ or 3) be a finite group satisfying the following conditions:*

(i) *$G(n,r)$ has no subgroup of index 2, and*

(ii) *$G(n,r)$ contains an involution α_n in the center of a 2-Sylow subgroup of $G(n,r)$ such that the centralizer $C_{G(n,r)}(\alpha_n)$ is isomorphic to $\tilde{H}(n,r)$. Then*

(1) *$G(n,1)$ is isomorphic to A_{4n} or A_{4n+1}, and*

(2) *$G(n,r)(r = 2$ or $3)$ is isomorphic to A_{4n+r}, except for the values of n and r listed below:*

n	r	Groups	References
1	2	A_6, PSL(2,7)	[5]
2	1	A_8, A_9, Hol(Z_2^3)	[1]
3	1	A_{12}, A_{13}, $Sp_6(2)$	[7]

where Hol(Z_2^3) *denotes the holomorph of an elementary abelian group of order 8.*

For small values $m \leq 15$, Theorem A was treated by K. A. Fowler, M. Suzuki [5], D. Held [1], [2], T. Kondo [4], and H. Yamaki [6], [7].

For the case $r = 2$ or 3, I have proved the following generalization of Theorem A.

AMS 1970 subject classifications. Primary 20B10; Secondary 20C30.

THEOREM B. *Let m be congruent to 2 or 3 mod 4, and $\tilde{z}$ be an (arbitrary) involution of A_m. Let G be a finite group with the following properties:*

(i) *G has no subgroup of index 2, and*

(ii) *G contains an involution z such that $C_G(z)$ is isomorphic to $C_{A_m}(\tilde{z})$.*

Then if $m \geq 7$, G is isomorphic to A_m.

For the remaining case $m \equiv 0$ or 1 mod 4, I have not yet obtained a similar result to Theorem B. The reason lies in just the point that we cannot find out any method to examine the fusion of involutions.

References

1. D. Held, *A characterization of the alternating groups of degrees eight and nine*, J. Algebra **7** (1967), 218–237. MR **36** #1530.

2. ———, *A characterization of some multiply transitive permutation groups.* I, Illinois J. Math. **13** (1969), 224–240. MR **39** #304.

3. T. Kondo, *On the alternating groups.* III, J. Algebra **14** (1970), 35–69.

4. ———, *A characterization of alternating groups of degree eleven*, Illinois J. Math. **13** (1969), 528–541. MR **40** #225.

5. M. Suzuki, *On finite groups containing an element of order four which commutes only with its powers*, Illinois J. Math. **3** (1959), 255–271. MR **21** #3486.

6. H. Yamaki, *A characterization of the alternating groups of degrees* 12, 13, 14, 15, J. Math. Soc. Japan **20** (1968), 673–694. MR **38** #1157.

7. ———, *A characterization of the finite simple groups* Sp(6,2), J. Math. Soc. Japan **21** (1969), 334–356.

INSTITUTE FOR ADVANCED STUDY

Character Tables and the Schur Index

Karl Kronstein

Let k be a field of characteristic 0, and χ an ordinary irreducible character of the finite group G; let K ($\supseteq k$) be a splitting field for χ. Let V be an irreducible $k[G]$-module which belongs to χ. Define the Schur index $m_k(\chi)$ to be the multiplicity of χ in $\chi_{V \otimes_k K}$.

The question we wish to consider is whether $m_k(\chi)$ can be computed from the character table together with certain information about the conjugate classes of G. In case k is the real field, the problem was solved for all groups G by Frobenius-Schur [1], who defined $c(\chi) = |G|^{-1} \sum_{g \in G} \chi(g^2)$, and proved that $c(\chi) = -1$ if and only if $m_k(\chi) = 2$.

Recently I was able to obtain a similar result [2] for hyperelementary groups and local fields. Let p, q be rational primes. G is called hyperelementary at p if it has a cyclic normal p-complement.

THEOREM. *Let k be the completion of the rationals at the prime q. Suppose that the irreducible character χ of the group G, which is hyperelementary at p, has the q-modular (Brauer) character φ as a constituent. Then*

$$m_k(\chi) = [k(\chi,\varphi):k(\chi)]$$

unless $p = q = 2$.

The Brauer character φ is "essentially" an ordinary character and the fact that it is a constituent of χ can be read from the character table. Also I have been able to find a character algorithm for the Schur index of hyperelementary groups over the completion of the rationals at 2.

At this point we would like to show that there can be no algorithm for the Schur

---

AMS 1970 *subject classifications.* Primary 20C15, 20C20; Secondary 16A26.

index which applies to arbitrary groups. In fact our result sets off the classical Frobenius-Schur formula by proving that when such an algorithm exists then $m_k(\chi) \leq 2$.

A one-one map $\sigma: G \to G$ is said to belong to Aut ccl G provided:
(1) σ permutes conjugate classes;
(2) If χ is a character of G (ordinary or modular), then χ^σ is a character of the same kind;
(3) $(C^m)^\sigma = (C^\sigma)^m$ for all conjugate classes C and m in Z. Here $C^m = \{x^m | x \in C\}$.

The following proposition is fairly obvious.

PROPOSITION. *If there is a σ in Aut ccl G for which $m_k(\chi) \neq m_k(\chi^\sigma)$, then $m_k(\chi)$ cannot be computed from the following information:*
(A) *character tables of G, ordinary and modular,*
(B) *the maps $C \to C^m$, for all m in Z.*

THEOREM. *Let k be an algebraic number field, or the completion of such a field at some prime. If $m_k(\chi)$ is computable from the character tables of G (ordinary and modular) and the maps $C \to C^m$, then $m_k(\chi) \leq 2$.*

PROOF. If false, there exists a group H, with ordinary irreducible ψ, and $m_k(\psi) \geq 3$. Let $G = H \times H$; $\chi(x, y) = \psi(x)\psi(y)$, and $(x, y)^\sigma = (x, y^{-1})$. Clearly σ belongs to Aut ccl G.

We claim that $m_k(\chi) \neq 1$, and that $m_k(\chi^\sigma) = 1$. Without loss of generality, assume that $k = k(\psi)$. Let A be the central simple algebra over k which belongs to ψ. Then $A \otimes_k A$ is central simple over k; belongs to χ; and cannot be the split algebra since because of the hypothesis on k, we have exp A = index $A = m_k(\psi) \geq 3$. However $A \otimes_k A^o$ belongs to χ^σ; here A^o is the opposite algebra. So $m_k(\chi^\sigma) = 1$. The theorem is proved.

REFERENCES

1. G. Frobenius and I. Schur. *Über die reellen Darstellungen der endlichen Gruppen*, S.-B. Preuss. Akad. Wiss. **1906**, 209-217.

2. K. Kronstein, *Representations over q-adic and q-modular fields*, J. Algebra (to appear).

UNIVERSITY OF NOTRE DAME

Restriction of Representations over Fields of Characteristic p

T. Y. Lam, I. Reiner, and D. Wigner[1]

1. **Introduction and statement of main results.** In this note we present some theorems on the restrictions of representations of finite groups over fields of characteristic p. These results will be formulated in the setting of relative Grothendieck rings, thus continuing works [1]–[6] by the first two authors.

To fix the notation, let Ω be a field of characteristic p, where $p \neq 0$. Let G be a finite group, and H a subgroup of G. By a "G-module" we shall always mean a left ΩG-module of finite dimension over Ω. Now form the free abelian group A on the symbols $[M]$, where M ranges over the representatives of a full set of isomorphism classes of G-modules. Let B be the subgroup of A generated by all expressions $[M] - [M'] - [M'']$, where

$$0 \to M' \to M \to M'' \to 0$$

ranges over all exact sequences of G-modules which are H-split. We then define the relative Grothendieck ring $a(G,H)$ as A/B, with the ring structure given by

$$[M][N] = [M \otimes_\Omega N],$$

where G acts diagonally on $M \otimes_\Omega N$. In the extreme case where $H = 1$, the ring $a(G,1)$ is just the ring of generalized Brauer characters of G (over Ω). On the other hand, when $H = G$, the ring $a(G,G)$ coincides with Green's representation ring, usually denoted by $a(G)$.

We can now state our first main result.

AMS 1970 *subject classifications.* Primary 20C05; Secondary 20J05, 18F30.

[1]This work was partially supported by a research contract with the NSF.

THEOREM 1. *Let Ω be an algebraically closed field of characteristic p. Let H be a normal subgroup of G such that $[G:H]$ is a power of p. Then the restriction of representations from G to H induces a ring isomorphism between $a(G,H)$ and the subring T of $a(H)$ spanned by all G-self-conjugate H-modules.*

The proof of Theorem 1 will be based on our second main result, Theorem 2 below, which is interesting in its own right. Before stating it explicitly, we need the following:

DEFINITION. Let M,N be G-modules, $M \neq 0$, and let $f: M \to N$ be an Ω-linear map. We call f a *local G-homomorphism* if there exists a nonzero G-submodule M_0 of M such that the restriction $f|M_0$ is a G-homomorphism of M_0 into N.

THEOREM 2 (LOCAL HOMOMORPHISM THEOREM). *Let H, G and Ω satisfy the hypotheses of Theorem 1, and let M,N be a pair of nonzero G-modules which are H-isomorphic. Then there exists an H-isomorphism $f: M_H \cong N_H$ which is a local G-homomorphism.*

By taking contragredient representations, we may deduce from Theorem 2 the following dual version (which is in fact equivalent to Theorem 2).

THEOREM 2'. *Let H,G,Ω,M,N be as in Theorem 2. Then there exist G-submodules $M' \subsetneq M$, $N' \subsetneq N$, and an H-isomorphism $g: M_H \cong N_H$, such that*
(i) $g(M') = N'$, *and*
(ii) *g induces a G-isomorphism $\bar{g}: M/M' \cong N/N'$.*

REMARK. In §3 we shall show that Theorem 2 implies Theorem 1, while §4 will be devoted to a proof of Theorem 2. It will be clear from these proofs that if Ω is *not* algebraically closed, but is instead a *finite* field of characteristic p, then Theorems 2 and 2' remain valid. The conclusion of Theorem 1 must be modified slightly, and states that in this case, the restriction map gives an isomorphism of $a(G,H)$ onto a subring of T.

The authors wish to thank Professors Simon Kochen and Kenneth Appel for helpful conversations dealing with the application of the Lefschetz Principle in §4.

2. **Corollaries and counterexamples.** We shall now deduce some corollaries of Theorems 1 and 2, and shall give examples to show that their hypotheses cannot be weakened. Throughout this section, the field Ω is assumed either finite or algebraically closed.

COROLLARY 1. *Suppose that $K \subset H \triangle G$, where $[H:K]$ is p-free and $[G:H]$ is a power of p. Then $a(G,K)$ is Z-free.*

PROOF. Since $[H:K]$ is p-free, an exact sequence of G-modules is H-split if and only if it is K-split, by a well-known analogue of Maschke's Theorem. Therefore $a(G,K) = a(G,H)$. But by Theorem 1 and the remark at the end of §1, $a(G,H)$ is isomorphic to a subring of the representation ring $a(H)$. Since $a(H)$ is Z-free by the

Krull-Schmidt Theorem, it now follows that $a(G,K)$ is also Z-free. This completes the proof.

COROLLARY 2. *If G has a normal p-complement C, then $a(G,K)$ is Z-free for any $K \triangle G$.*

PROOF. Set $H = K \cdot C$. Then the triple $K \subset H \triangle G$ satisfies the hypotheses of Corollary 1, so we may conclude that $a(G,K)$ is Z-free.

We next record a consequence of Theorem 2.

COROLLARY 3. *Let $H \triangle G$, where $[G:H] = $ power of p, and let Ω be algebraically closed of characteristic p. Let M,N be a pair of nonzero H-isomorphic G-modules. Then there exist G-composition series*

$$0 \subsetneq M_0 \subsetneq M_1 \subsetneq \ldots \subsetneq M_n = M,$$
$$0 \subsetneq N_0 \subsetneq N_1 \subsetneq \ldots \subsetneq N_n = N,$$

such that for $0 \leq i \leq n - 1$,

$$\frac{M_{i+1}}{M_i} \cong \frac{N_{i+1}}{N_i} \quad \text{as } G\text{-modules.}$$

In particular, if M is G-irreducible, then $M \cong N$.

PROOF. By Theorem 2, there exist a nonzero G-submodule $M_0 \subset M$, and an H-isomorphism $f: M_H \cong N_H$, such that $f|M_0$ is a G-homomorphism. Replacing M_0 by one of its G-irreducible submodules if necessary, we may assume that M_0 is itself G-irreducible. Then setting $N_0 = f(M_0) \subset N$, we see that N_0 is a G-irreducible submodule of N, and that f induces an H-isomorphism $\bar{f}: (M/M_0)_H \cong (N/N_0)_H$. We now invoke an inductive hypothesis, and complete the proof by pulling back suitable G-composition series from M/M_0 and N/N_0.

REMARKS. 1. As pointed out earlier, Theorem 2 and Corollaries 1–3 remain valid if Ω is any finite field of characteristic p.

2. Since $H \triangle G$ and $[G:H]$ is a power of p, it is clear that the p-regular elements of G all lie in H. However, the Brauer character μ afforded by a G-module M is a function defined only on p-regular elements of G, and μ determines the composition factors of M. Therefore the restriction map $\text{res}: a(G,1) \to a(H,1)$ is a ring monomorphism. Hence, if a pair of G-modules M,N are H-isomorphic (as in the hypothesis of Corollary 3), it is no surprise that M,N have the same G-composition factors, apart from order of occurrence. The corollary asserts a much stronger result, however, namely that M,N possess G-composition series with *matching* composition factors in the *same* order of occurrence.

Let us now give examples to show that the hypotheses of Theorems 1 and 2 cannot be weakened. To begin with, if $H \triangle G$ but $[G:H]$ is not a power of p, then $a(G,H)$ is "much bigger" than $a(H)$. For example (see [2]), if $G = H \times A$ (direct product), then

$$a(G,H) \cong a(H) \otimes_Z a(A,1).$$

Thus in Theorem 1 we cannot relax the hypothesis that $[G:H]$ be a power of p.

On the other hand, if H is not normal in G, then $a(G,H)$ may still be "larger" than $a(H)$, even if $[G:H]$ is a power of p. For example, take $p = 5$, $H = S_4$, $G = S_5$. Then H has five 5-regular conjugacy classes, while G itself has six such classes, the element $(123)(45)$ giving the extra class. Since $a(G,H) = a(G,1)$ because H is a p'-group, we see that $a(G,H)$ has Z-rank 6. On the other hand $a(H) = a(H,1)$, which is of Z-rank 5. Therefore $a(G,H)$ cannot be isomorphic to a subgroup of $a(H)$.

We shall give one final example in which G is itself a p-group, H is a nonnormal subgroup of G, and the conclusion of Theorem 2 does not hold. Furthermore, in this particular example, the conclusion of Theorem 1 is valid! In fact, let $p = 2$ and let G be the dihedral group of order 8, with the presentation

$$G = \langle x,y | x^2 = y^4 = 1, xyx = y^3 \rangle.$$

Let $H = \langle x \rangle$, and let Ω be any field of characteristic 2. Set $\xi = x - 1$, $\eta = y - 1$ in the group algebra ΩG. It is easily checked that all relations between ξ and η are consequences of the relations

(1) $$\xi^2 = \eta^4 = 0, \quad \xi\eta\xi + \xi\eta + \eta\xi = \eta^3 + \eta^2.$$

Now let

$$M = \Omega e_1 \oplus \Omega e_2 \oplus \Omega e_3, \quad N = \Omega e'_1 \oplus \Omega e'_2 \oplus \Omega e'_3,$$

where the action of G on M,N is given by

$$\begin{array}{llll} \xi e_1 = e_2 & \eta e_1 = e_2 & \xi e'_1 = e'_2 & \eta e'_1 = e'_2 \\ \xi e_2 = 0 & \eta e_2 = 0 & \xi e'_2 = 0 & \eta e'_2 = e'_3 \\ \xi e_3 = 0 & \eta e_3 = e_1 & \xi e'_3 = 0 & \eta e'_3 = 0. \end{array}$$

Using (1), it is a trivial exercise to check that these formulas give a *well-defined* action of G on M and N. (In fact, these correspond to the two faithful 3-dimensional representations of the dihedral group G arising from the two inequivalent imbeddings of G into the simple group $GL_3(\mathbf{F}_2)$ of order 168.)

It is clear that $M_H \cong N_H$, but M,N are not G-isomorphic. We claim that there does not exist any H-isomorphism $f: M \cong N$ which is simultaneously a local G-homomorphism. Suppose to the contrary that such an f exists. Since Ωe_2 is the unique irreducible G-submodule of M, and $\Omega e'_3$ is the unique irreducible G-submodule of N, the local G-homomorphism f must map Ωe_2 onto $\Omega e'_3$. But since f is also an H-isomorphism, we obtain

$$f(\Omega e_2) = f(\xi M) = \xi f(M) = \xi N = \Omega e'_2,$$

a contradiction. This shows that the conclusion of Theorem 2 is *not* valid in this case. On the other hand, Theorem 2.1 of [5] implies that

$$\text{res}: a(G,H) \cong T = a(H),$$

so the conclusion of Theorem 1 is true here.

3. **Theorem 2 implies Theorem 1.** Suppose that the hypotheses of Theorem 1 are satisfied, and let φ denote the restriction map from $a(G,H)$ to $a(H)$. By Theorem 6.12 of [**5**], the image of φ is precisely the subring T of $a(H)$ spanned by all G-self-conjugate H-modules. Thus, to prove Theorem 1, we need only show that φ is monic, that is, for G-modules M and N, if $M_H \cong N_H$ then $[M] = [N]$ in $a(G,H)$. We shall deduce this from Theorem 2, using induction on $\dim_\Omega M$. There is nothing to prove when $M = 0$, so assume now that $M \neq 0$. By Theorem 2, we may pick an H-isomorphism $f: M_H \cong N_H$ which is a local G-homomorphism, that is, there exists a nonzero G-submodule M_0 of M such that $f|M_0$ is a G-homomorphism of M_0 into N. Consequently there exists a commutative diagram

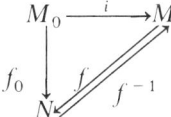

where $f_0 = f|M_0$. Both f_0 and i are G-monomorphisms, while f and f^{-1} are H-homomorphisms. By the Pushout Lemma (see [**2**, 3.1]), this implies that the following equation holds true in $a(G,H)$:

$$[M] - [M/M_0] = [N] - [N/N_0],$$

where $N_0 = f(M_0)$. But f induces an H-isomorphism $(M/M_0)_H \cong (N/N_0)_H$, so by the inductive hypothesis, $[M/M_0] = [N/N_0]$ in $a(G,H)$. Therefore $[M] = [N]$ in $a(G,H)$, as desired, and Theorem 1 is proved.

4. **Proof of Theorem 2.** The hypotheses and notations of Theorem 2 will remain in force throughout this section. Let P be a p-Sylow subgroup of G. Since $[G:H]$ is a power of p, we have $G = HP$. We shall construct an Ω-linear map $f: M \to N$ satisfying

(1) f is an H-isomorphism, and

(2) there exists a nonzero P-fixed point $m \in M$ such that $f(m)$ is a P-fixed point of N.

From these conditions we may at once deduce that f is a local G-homomorphism. In fact, let

$$M_0 = \Omega G \cdot m = \Omega H \cdot m,$$

a nonzero G-submodule of M. For any $g \in G$ and $m_0 \in M_0$, write

$$g = hx, h \in H, x \in P, \qquad m_0 = \xi m, \xi \in \Omega H.$$

Then
$$\begin{aligned}
f(gm_0) &= f(hx \cdot \xi m) \\
&= f(h \cdot x\xi x^{-1} \cdot m) \quad \text{(since } m \text{ is } P\text{-fixed)}, \\
&= h \cdot x\xi x^{-1} \cdot f(m) \quad \text{(since } h \cdot x\xi x^{-1} \in \Omega H), \\
&= hx \cdot \xi \cdot f(m) \quad \text{(since } f(m) \text{ is } P\text{-fixed)}, \\
&= g \cdot f(m_0) \quad \text{(since } \xi \in \Omega H).
\end{aligned}$$

Thus $f|M_0$ is a G-homomorphism, as desired. Hence Theorem 2 will be established once we prove the existence of a map f satisfying (1) and (2).

Let us identify M and N as H-modules. The existence of the map f satisfying (1) and (2) is equivalent to the following statement.

PROPOSITION. *Let $\lambda_i: G \to GL_n(\Omega)$ ($i = 1,2$) be two matrix representations of G, such that $\lambda_1(h) = \lambda_2(h)$ for every $h \in H$. Let*

$$H_0 = \lambda_1(H) = \lambda_2(H), \qquad P_1 = \lambda_1(P), \qquad P_2 = \lambda_2(P).$$

Then there exists a matrix $\rho \in GL_n(\Omega)$ such that

(1') *ρ centralizes H_0, and*

(2') *the matrix groups P_2 and $\rho^{-1}P_1\rho$ have a common nonzero fixed vector $v = (v_1, \ldots, v_n) \in \Omega^n$.*

PROOF. To prove the proposition, we first make a substantial reduction of the problem by appealing to a theorem of mathematical logic, namely the "Lefschetz Principle." This powerful theorem states that if X is an "elementary statement" for fields of a fixed nonzero characteristic p, and if X is true for one *specific* algebraically closed field Ω_0 of characteristic p, then X automatically holds true for *every* algebraically closed field Ω of characteristic p.[2] (As a reference for the Lefschetz Principle, see [7], especially Chapter 4 on the notion of "completeness.")

Now the hypotheses of the proposition involve only matrix representations over the field Ω, while the conclusion involves only properties such as commutativity of matrices and existence of a nonzero fixed vector. The proposition is therefore itself an elementary statement about the field Ω. By the Lefschetz Principle, it thus suffices to prove the proposition for the special case where the underlying field is the algebraic closure Ω_0 of the prime field $GF(p)$. Let us proceed with the proof for this special case.

Since both P_1 and P_2 are finite subgroups of $GL_n(\Omega_0)$, it follows at once that there exists a *finite* field $\Omega' \subset \Omega_0$ such that $P_i \subset GL_n(\Omega')$, $i = 1,2$. Now P_1 normalizes H_0, since for $x \in P$, $h \in H$, we have

$$\lambda_1(x) \cdot \lambda_1(h) \cdot \lambda_1(x)^{-1} = \lambda_1(xhx^{-1}) \in H_0.$$

Thus the elements of P_1 act on H_0 by conjugation, and there is a homomorphism

$$t_1: P_1 \to \text{Aut}(H_0),$$

---

[2] In the language of [7, Theorem 4.14]: "The concept of an algebraically closed field of specified characteristic is complete."

where $\text{Aut}(H_0)$ is the automorphism group of H_0. Likewise, P_2 normalizes H_0, and there is a homomorphism $t_2: P_2 \to \text{Aut}(H_0)$. Furthermore, since $\lambda_1 = \lambda_2$ on H, we have

$$\lambda_1(xhx^{-1}) = \lambda_2(xhx^{-1}), \qquad x \in P, \quad h \in H.$$

This shows that for each $x \in P$, the matrix $\lambda_2(x)^{-1} \cdot \lambda_1(x)$ centralizes H_0, or equivalently,

(3) $$t_1(\lambda_1(x)) = t_2(\lambda_2(x)), \qquad x \in P.$$

Now let $U = \langle P_1, P_2 \rangle$ be the subgroup of $GL_n(\Omega')$ generated by P_1 and P_2. Since Ω' is a finite field, the group U is finite. Furthermore, U normalizes H_0 since both P_1 and P_2 normalize H_0. Thus there exists a homomorphism $t: U \to \text{Aut}(H_0)$ making the following diagram commute:

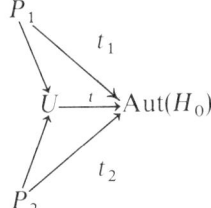

(The maps $P_i \to U$ are inclusion maps.) Equation (3) tells us that t_1 and t_2 have the same image in $\text{Aut}(H_0)$. Since these images generate the image of t, it follows at once that the image of t in $\text{Aut}(H_0)$ coincides with the image of t_1.

Next, we observe that P_1 and P_2 are p-subgroups of the finite group U, and so there exist p-Sylow subgroups Q_1, Q_2 of U such that $P_i \subset Q_i \subset U$, $i = 1,2,\ldots$. Surely $Q_2 = u^{-1} Q_1 u$ for some element $u \in U$. By the remarks at the end of the preceding paragraph, we may write

$$t(u) = t_1(\lambda_1(x)) \quad \text{for some } x \in P.$$

Hence $u = \lambda_1(x) \cdot \rho$, where ρ is a matrix in $GL_n(\Omega')$ centralizing H_0. We have

$$Q_2 = u^{-1} Q_1 u = \rho^{-1} \cdot \lambda_1(x)^{-1} \cdot Q_1 \cdot \lambda_1(x) \cdot \rho = \rho^{-1} Q_1 \rho,$$

since $\lambda_1(x) \in P_1 \subset Q_1$. Then $\rho^{-1} P_1 \rho \subset \rho^{-1} Q_1 \rho = Q_2$, so both of the matrix groups P_2 and $\rho^{-1} P_1 \rho$ are subgroups of the group Q_2. However, Q_2 is a p-group acting on a vector space $(\Omega')^n$ over a field Ω' of characteristic p, and so there exists a nonzero vector $v \in (\Omega')^n$ fixed by Q_2. But then both P_2 and $\rho^{-1} P_1 \rho$ fix v, as desired, and the proposition is proved.

The preceding proof also establishes the proposition when Ω is any *finite* field. It seems likely that the proposition, and Theorem 2 as well, are valid for an arbitrary field Ω of characteristic p. We cannot prove this at present, however.

References

1. T. Y. Lam and I. Reiner, *Relative Grothendieck groups*, J. Algebra **11** (1969), 213–242. MR **38** #4574

2. ———, *Reduction theorems for relative Grothendieck rings*, Trans. Amer. Math. Soc. **142** (1969), 421–435. MR **39** #7003.

3. ———, *Relative Grothendieck rings*, Bull. Amer. Math. Soc. **75** (1969), 496–498. MR **39** #326.

4. ———, *Finite generation of Grothendieck groups relative to cyclic subgroups*, Proc. Amer. Math. Soc. **23** (1969), 481–489. MR **40** #1495.

5. ———, *Restriction maps on relative Grothendieck rings*, J. Algebra **14** (1970), 260–298.

6. ———, *An excision theorem for Grothendieck rings*, Math. Z. **115** (1970), 153–164.

7. A. Robinson, *Introduction to model theory and to the metamathematics of algebra*, North-Holland, Amsterdam, 1963. MR **27** #3533.

University of California, Berkeley

University of Illinois, Urbana

Stanford University

On the Suzuki and Conway Groups

J. H. Lindsey II

As in [2], all six by six even permutation matrices, all unimodular, diagonal, six by six matrices of order 3, and $U \oplus \bar{U}$, where

$$U = \begin{pmatrix} 1 & 1 & 1 \\ 1 & \omega & \bar{\omega} \\ 1 & \bar{\omega} & \omega \end{pmatrix} \bigg/ \sqrt{-3}$$

and ω is a primitive third root of unity, generate a complex linear group P isomorphic to a proper central extension of Z_6 by $PSU_4(3)$. This fact was used to construct the following generators of a 12-dimensional, unitary, complex linear group S isomorphic to a proper central extension of Z_6 by the simple Suzuki group of order $2^{13}3^7 5^2 7.11.13$:

$$M_1 = U \oplus \bar{U} \oplus \bar{U} \oplus U$$
$$M_2 = \omega I_3 \oplus \bar{\omega} I_3 \oplus \omega I_3 \oplus \bar{\omega} I_3$$
$$M_3 = (1, 2, 3)(1', 2', 3')(4', 5', 6')$$
$$M_4 = (4, 5, 6)(1', 2', 3')(4', 6', 5')$$
$$M_5 = (1, 2, 6)(1', 3', 4')(2', 6', 5')$$
$$M_6 = (1, 2, 2')(5, 4', 4)(6, 5', 6')$$

where the last four matrices are permutation matrices acting on the 12 variables $x_1, \ldots, x_6, x_{1'}, \ldots, x_{6'}$ corresponding to the given permutations.

The first five generators generate a reducible linear group H isomorphic to a proper central extension of $Z_6 \times Z_3$ by $PSU_4(3)$. The last generator was constructed in $C(M_2)$ to play the same role as M_5 played in the centralizer of $\omega I_6 \oplus \bar{\omega} I_6$ (conjugate to M_2 in S).

AMS 1970 *subject classifications*. Primary 20G20.

The last four generators generate M_{11} and permute the eleven partitions $\{1,\ldots,6, 1',\ldots,6'\} = S_i \cup S'_i, i = 1,\ldots, 11$ where

$$S_1 = \{1, 2, 3, 4, 5, 6\}$$
$$S_2 = \{1, 2, 3, 1', 2', 3'\}$$
$$S_3 = \{1, 2, 4, 3', 4', 5'\}$$
$$S_4 = \{1, 2, 5, 1', 4', 6'\}$$
$$S_5 = \{1, 3, 4, 2', 4', 6'\}$$
$$S_6 = \{1, 3, 5, 3', 5', 6'\}$$
$$S_7 = \{1, 4, 5, 1', 2', 5'\}$$
$$S_8 = \{2, 3, 4, 1', 5', 6'\}$$
$$S_9 = \{2, 3, 5, 2', 4', 5'\}$$
$$S_{10} = \{2, 4, 5, 2', 3', 6'\}$$
$$S_{11} = \{3, 4, 5, 1', 3', 4'\}.$$

The last five generators generate a monomial group $3.3^5.M_{11}$ where $A.B$ denotes an extension of the group A by the group B. The center 3. does not have an M_{11} complement in the elementary abelian group 3.3^5.

Our 12-dimensional group S preserves the lattice $\mathscr{L}$ of $(a_1,\ldots, a_{6'})$ satisfying the following:

I. $a_i \in Z[\omega]$, all $i = 1,\ldots, 6'$;
II. $a_i - a_j \in (-3)^{1/2}(Z[\omega])$, all i, j;
III. $\sum_{j \in S} a_j \in 3(Z[\omega])$ for $S = S_i$ and S'_i and all i;
IV. $3a_1 + \sum_{j=1}^{6'} a_j \in 3(-3)^{1/2}(Z[\omega])$.

Then $\{2^{1/2}(\text{Re } a_1, \text{Im } a_1, \text{Re } a_2, \text{Im } a_2,\ldots, \text{Im } a_{6'})/3 \mid (a_1,\ldots, a_{6'}) \in \mathscr{L}\}$ is a unimodular lattice (one point per unit volume) with every squared length an even integer greater than two, and must be the Leech lattice by the characterization in [1].

We define N_7 to be the Z-linear function $(a_1,\ldots, a_{6'})N_7 = (\bar{a}_{6'}, a_1, a_3, \bar{a}_{2'}, a_6, \bar{a}_{3'}, \bar{a}_4, a_{1'}, \bar{a}_5, a_{4'}, a_{5'}, \bar{a}_2)$. Then N_7 preserves $\mathscr{L}$ and along with the linear functions given by $M_1,\ldots, M_6$ generates C, the 24-dimensional Conway group. Also, $M_2,\ldots, M_6, N_7$ generate $3^6.2.M_{12}$.

The 12-dimensional group S is transitive on the nearest and second nearest points of $\mathscr{L}$ to the origin. Then $C = SF_1 = SF_2$ where F_i is the subgroup of C fixing the ith nearest lattice point. The subgroup of S fixing a nearest lattice point seems likely to be isomorphic to $PSU_5(2)$.

There are further similarities between the linear groups P, S, and C in the centralizer of an involution. This centralizer seems likely to be the quaternions tensored with themselves n times extended by a subgroup of index two in $O_{2n}^{(-)n}(2)$ for $n = 2, 3, 4$ for P, S, and C, respectively.

These constructions make it easy to show that the simple Conway group is complete and the simple Suzuki group has outer automorphism group of order 2 (coming from complex conjugation of S).

It may be difficult or impossible to generalize this construction to a 48-dimensional group F. The two components of the reducible subgroup corresponding to P or S of S or C, respectively, were related by nontrivial outer automorphisms of the projective groups, $PSU_4(3)$ and $S/Z(S)$, respectively. As the Conway group is complete, this twist is no longer possible and we would have to assume that the 48-dimensional group contained

$$C \oplus C = \left\{ c \oplus c = \begin{pmatrix} c & 0 \\ 0 & c \end{pmatrix} \,\bigg|\, c \in C \right\}.$$

This contains $S \oplus \bar{S} \oplus S \oplus \bar{S}$ and $T = \omega I_{12} \oplus \bar{\omega} I_{12} \oplus \omega I_{12} \oplus \bar{\omega} I_{12}$. As $S \oplus S$ is a subgroup of the representation of $C_F(T)$ on the ω-space of T, $C_F(T)$ cannot be the Conway group. (C contained $S \oplus \bar{S}$, not $S \oplus S$.) Therefore, things cannot be completely analogous to the earlier constructions of S and C.

References

1. J. H. Conway, *A group of order* 8,315,553,613,086,720,000, Bull. London Math. Soc. **1** (1969), 79–88. MR **40** #1470.

2. J. H. Lindsey II, *Linear groups of degree* 6 *and the Hall-Janko group*, Proc. Sympos. Theory of Groups (Harvard University, Cambridge, Mass., 1968) Benjamin, New York, 1969, pp. 97–100. MR **39** #1538.

NORTHERN ILLINOIS UNIVERSITY

Matrix Questions and the Brauer-Thrall Conjectures on Algebras with an Infinite Number of Indecomposable Representations

L. A. Nazarova and A. V. Roiter

It is well known that a finite-dimensional semisimple algebra has a finite number of indecomposable representations. But nonsemisimple algebras may have an infinite number of indecomposable representations. This infinity may occur for two reasons:

(1) the algebra may have indecomposable representations of arbitrarily high dimension, and

(2) (if the base field is infinite) there may exist infinitely many indecomposable representations of some fixed dimension.

Following Brauer (see [1]), we say that an algebra is of finite type if it has a finite number of indecomposable representations and of infinite type otherwise. Furthermore, we say that an algebra is of bounded type if the set of dimensions of its indecomposable representations is bounded (and of unbounded type otherwise). We say that an algebra is of strongly unbounded type if there exists an infinite number of dimensions, each admitting infinitely many indecomposable representations.

A priori we might assume (in the case of an infinite field) the existence of algebras of the following types:

(1) algebras of finite type,

(2) algebras of infinite, but of bounded, type,

(3) algebras of infinite type having a finite number of indecomposable representations of every fixed dimension,

(4) algebras of infinite type having an infinite number of indecomposable representations for only a finite number of dimensions,

(5) algebras of strongly unbounded type.

AMS 1970 *subject classifications*. Primary 16A64, 16A46, 15A21.

While algebras of types (1) and (5) are very easy to construct,[1] an investigation reveals that algebras of types (2), (3), and (4) apparently do not exist.[2] This was conjectured by Brauer and Thrall (see [1]). More precisely, Brauer and Thrall formulated two conjectures:

I. Any algebra of bounded type is of finite type.

II. If the field is infinite, any algebra of unbounded type is of strongly unbounded type.

The first of these conjectures (which implies the nonexistence of algebras of type (2)) was proved by one of the present authors in [2].[3]

We remark that the proof of the first conjecture is based on rather general concepts and is technically very simple. The need to study the structure of algebras with an infinite number of indecomposable representations and the representations themselves does not arise. Conjecture I also admits various generalizations. In particular we have

PROPOSITION 1. *If a set of finitely generated modules over an Artinian ring which is closed with respect to the formation of a direct sum and a factor module contains infinitely many indecomposable modules, then this set contains indecomposable modules of arbitrarily high dimension.*

We also have a dual Proposition 1*, obtained from Proposition 1 by replacing "factor module" by "submodule."

The situation is different with the second Brauer-Thrall conjecture. We first observe that it clearly does not admit the generalizations analogous to Propositions 1 and 1*. Indeed, by the methods of [2] it is not difficult to construct an example of a finite-dimensional algebra Λ over an infinite field and a set of Λ-modules which is closed with respect to the formation of a direct sum and a factor module and which contains an infinite number of indecomposable modules, but a finite number of indecomposable modules of every fixed dimension.

This shows that the ideas behind Conjecture II are much more specific and related to the actual structure of infinite sets of indecomposable modules. So to study these problems we use the technique of matrix questions, which was developed in [5], [6] to classify infinite sets of indecomposable modules over rings of a special form.

---

[1] An example of a nonsemisimple algebra of finite type is the two-dimensional algebra with basis 1, τ; $\tau^2 = 0$. An example of an algebra of strongly unbounded type is the three-dimensional algebra generated by $1, \tau_1, \tau_2; \tau_1\tau_2 = \tau_2\tau_1 = \tau_1^2 = \tau_2^2 = 0$, and this algebra has infinitely many indecomposable representations of every even dimension.

[2] Here we mean the field to be infinite. For a finite field it is clear that only cases (1) and (3) are possible.

[3] This was proved by Yoshii [3] under the assumption that the field be algebraically closed and the square of the radical of the algebra be zero, and by Curtis and Jans [4] under the assumption that any irreducible module occur in the socle of the algebra at most once.

Using this technique, we succeeded in settling the Brauer-Thrall Conjecture II under the assumption that the base field be algebraically closed. Thus we can state

PROPOSITION 2. *If a finite-dimensional algebra Λ over an algebraically closed field k has infinitely many indecomposable representations, then there exist infinitely many dimensions, each admitting infinitely many indecomposable representations.*

Furthermore, using the usual methods of extending the base field, it is not difficult to replace the requirement that the base field be algebraically closed by the substantially weaker requirement that the factor algebra $\bar{\Lambda} = \Lambda/\mathscr{R}$ be separable, $\mathscr{R}$ being the radical of Λ. In particular, the assertion is valid for algebras over a perfect field (hence for algebras over fields of characteristic zero).

As to algebras with nonseparable factor algebra, it appears that the difficulties which arise are not fundamental ones and can be handled by the same methods.

As already stated above, the basic method for proving Proposition 2 is the consideration of matrix questions, which, roughly speaking, refers to the form to which a set of matrices can be reduced by means of a set of "admissible" transformations.

Classical examples of matrix questions are the (trivial) question of the reduction of a single matrix by elementary transformations and the question of the reduction of two matrices by simultaneous elementary transformations.

Another example, arising in the theory of integral representations, is the following problem: the reduction of n given matrices (with the same number of rows), with any simultaneous elementary transformations permitted on the rows, and any elementary transformations permitted on the columns of each matrix. (This problem admits a solution for $n \leq 4$.)

Examples of matrix problems can also be found in [5], [6], [7].

We resist the temptation to give a rigorous general definition of a matrix question, but it does seem that after the suitable formal language is set forth (as a result of establishing a reasonable level of generality), the development of a "general theory of matrix questions" will turn out to be meaningful and useful for various applications.

In the proof of Proposition 2, the decisive role is played by matrix questions of the following type, which we call linear matrix questions. Let $\mathfrak{A}$ be a finite partially ordered set: $\mathfrak{A} = \{a_1, \ldots, a_n\}$. Any collection of matrices $A_1, \ldots, A_n$ (with entries from some field) will be called an object of the matrix question $\mathfrak{A}$ if all of the matrices $A_1, \ldots, A_n$ have the same number of rows.

In this situation we introduce the admissible transformations in the following way:

(1) simultaneous elementary transformations on the rows of all of the matrices $A_1, \ldots, A_n$,

(2) elementary transformations on the columns of any matrix A_i,

(3) when $a_i < a_j$, any column of A_i may be added to any column of A_j.

Two objects are called equivalent if each of the corresponding sets of matrices can be transformed into the other by means of (1)–(3).

For linear matrix questions we can introduce in a natural way the notions of indecomposable object, questions of finite and infinite types, questions of strongly unbounded type, etc.

PROPOSITION 3. *Any linear matrix question of infinite type over an algebraically closed field is of strongly unbounded type.*

In order to prove Proposition 3, it is necessary to develop a special algorithm for the solution of linear matrix questions which reduces the matter to the combinatorial analysis of partially ordered sets.

Let a be a maximal element of the partially ordered set $\mathfrak{A}$. We construct a new partially ordered set $\mathfrak{A}_a$ as follows: $\mathfrak{A}_a$ is the union of two subsets, namely the set $\mathfrak{A}$ without the element a and the set of all unordered pairs (b,c) of elements of $\mathfrak{A}$ such that b and c are noncomparable to one another and noncomparable to c. The ordering on $\mathfrak{A}_a$ is obtained by restricting the following order, which can be introduced on the subsets of any partially ordered set: let M and N be subsets of $\mathfrak{A}$, $M = (m_1, \ldots, m_k)$, $N = (n_1, \ldots, n)$; set $M \leq N$ if for any $m_i \in M$ there exists $n_j \in N$ such that $m_i \leq n_j$.

We can now state

PROPOSITION 4. *The matrix questions corresponding to the sets $\mathfrak{A}$ and $\mathfrak{A}_a$ are of the same type if $\mathfrak{A}$ has no more than three noncomparable elements.*

Thus the proof of Proposition 3 reduces to a rather tedious, purely combinatorial computation in finite partially ordered sets, involving the repeated application of the operation of passing from the set $\mathfrak{A}$ to the set $\mathfrak{A}_a$. The algorithm also allows us to construct all indecomposable objects for any linear matrix question.

We now give a brief indication of the proof of Proposition 2.

Suppose that Λ is an algebra of infinite type (over an algebraically closed field). We have to prove that Λ is an algebra of strongly unbounded type. We may clearly assume that Proposition 2 is valid for algebras of smaller dimension. Let W be some two-sided ideal of Λ. We may choose it such that Ann $W = \{\lambda; W\lambda = 0\} \neq 0$. Put $\Lambda' = \Lambda/\text{Ann } W$. We may clearly assume that Λ' is of finite type and that $A_1, \ldots, A_n$ are all of its indecomposable representations. Now suppose that M is any representation module of Λ. Put $M' = \{m \in M; m \text{ Ann } W = 0\}$, $M'' = M/M'$. It is obvious that M' can be regarded as a Λ'-module, and M'' as a module over $\Lambda'' = \Lambda/W$, and Λ'' is a full matrix ring possessing a unique indecomposable representation $\mathcal{U}$.

Thus to any M there corresponds an element of $\text{Ext}(\mathcal{U}^{(m)}, A_1^{(m_1)} \oplus A_2^{(m_2)} \oplus \ldots \oplus A_n^{(m_n)})$, which can be regarded as a set of matrices over the base field. The matrix question obtained as a result is not, in general, linear in the sense of the definition given above, but can be nontrivially reduced to several linear questions.

Proposition 2 now follows from Proposition 3.

References

1. J. P. Jans, *On the indecomposable representations of algebras*, Ann. of Math. (2) **66** (1957), 418–429. MR **19**, 526.

2. A. V. Roĭter, *Unbounded dimensionality of indecomposable representations of an algebra with an infinite number of indecomposable representations*, Izv. Akad. Nauk SSSR Ser. Mat. **32** (1968), 1275–1283 = Math. USSR Izv. **2** (1968), 1223–1230. MR **39** #253.

3. T. Yoshii, *Note on algebras of bounded representation type*, Proc. Japan Acad. **32** (1956), 441–445. MR **18**, 462.

4. C. W. Curtis and J. P. Jans, *On algebras with a finite number of indecomposable modules*, Trans. Amer. Math. Soc. **114** (1965), 122–132. MR **31** #2270.

5. L. A. Nazarova, *Representation of a tetrad*, Izv. Akad. Nauk SSSR Ser. Mat. **31** (1967), 1361–1378 = Math. USSR Izv. **1** (1967), 1305–1322. MR **36** #6400.

6. L. A. Nazarova and A. V. Roĭter, *Finitely generated modules over a dyad of two local Dedekind rings, and finite groups with an abelian normal divisor of index p*, Izv. Akad. Nauk SSSR Ser. Mat. **33** (1969), 65–89 = Math. USSR Izv. **3** (1969), 65–86.

7. Ju. A. Drozd and A. V. Roĭter, *Commutative rings with a finite number of indecomposable integral representations*, Izv. Akad. Nauk SSSR Ser. Mat. **31** (1967), 783–798 = Math. USSR Izv. **1** (1967), 757–772. MR **36** #3768.

ACADEMY OF SCIENCE OF UKRAINIAN SSR

Group Rings of Infinite Groups

D. S. Passman

Let $K[G]$ denote the group ring of a group G over a field K. If G is finite, then $K[G]$ can be studied using trace functions and the fairly strong structure theorems for finite dimensional algebras. If G is infinite, then these tools are no longer available and the study of $K[G]$ is in a much more primitive state. At present we merely view $K[G]$ as an interesting example of an algebra and we subject it to ring theoretic scrutiny. I will give a brief survey of some aspects of this study.

There is an interesting technique which pervades much of this subject. Let $\Delta = \Delta(G) = \{x \in G \mid [G:C(x)] < \infty\}$ so that Δ is a characteristic subgroup of G, its so-called F.C. subgroup. We define a projection map $\theta: K[G] \to K[\Delta]$ by

$$\alpha = \sum_{x \in G} k_x x \to \theta(\alpha) = \sum_{x \in \Delta} k_x x.$$

Then θ is clearly a K-linear map but it is certainly not a ring homomorphism in general. Nevertheless we have the following

Δ-LEMMA (PASSMAN [6]). *Let $\alpha, \beta \in K[G]$ and suppose that for all $x \in G$ we have $\alpha x \beta = 0$. Then $\theta(\alpha)\theta(\beta) = 0$.*

We will call any consequence of this lemma and its numerous variants a Δ-theorem. These results are generally proved in two steps. The first is a reduction from $K[G]$ to $K[\Delta]$ and the second is the pertinent study of $K[\Delta]$. An indication of the basically simple nature of the group Δ is seen in the following. Namely, the set Δ^+ of all elements of finite order in Δ is in fact a locally finite characteristic subgroup of Δ and Δ/Δ^+ is torsion free abelian.

Prime rings. A ring R is said to be prime if for any two ideals I_1, I_2 in R, $I_1 I_2 = 0$

AMS 1970 *subject classifications*. Primary 16–02, 16A26; Secondary 16A12, 16A20, 16A38.

implies $I_1 = 0$ or $I_2 = 0$. The question of when $K[G]$ is prime has been completely settled.

Δ-THEOREM 1 (CONNELL [3]). *The following are equivalent*:
(i) $K[G]$ *is prime*.
(ii) $\Delta(G)$ *is torsion free abelian*.
(iii) G *has no nonidentity finite normal subgroup*.

A remarkable but immediate consequence of this is

COROLLARY (CONNELL [3]). $K[G]$ *is Artinian if and only if G is finite*.

Necessary and sufficient conditions for $K[G]$ to be (right) Noetherian are apparently not known at present.

Semiprime rings. An ideal P of a ring R is said to be prime if R/P is a prime ring. R is said to be semiprime if the intersection of all prime ideals of R is zero. Thus R is semiprime if and only if it is a subdirect product of prime rings. An elementary lemma in ring theory states that R is semiprime if and only if R has no nonzero nilpotent ideals.

THEOREM 2. *If K has characteristic 0, then $K[G]$ is semiprime*.

Δ-THEOREM 3 (PASSMAN [6]). *Let K have characteristic $p > 0$. Then the following are equivalent*:
(i) $K[G]$ *is semiprime*.
(ii) $\Delta(G)$ *is a p'-group*.
(iii) G *has no finite normal subgroup whose order is divisible by p*.

Thus the question of when $K[G]$ is semiprime has been completely settled. Let $NK[G]$ denote the nilpotent radical of $K[G]$, that is the sum of all nilpotent ideals in $K[G]$.

Δ-THEOREM 4 (PASSMAN [9]). *Let K have characteristic $p > 0$. Then*
(i) $NK[G] = (NK[\Delta])K[G]$.
(ii) $NK[G]$ *is nilpotent if and only if Δ contains only finitely many elements whose order is a power of p*.

Polynomial identity rings. Let $K[\zeta_{1,2},\ldots]$ be the polynomial ring over K in the noncommuting indeterminates $\zeta_1, \zeta_2, \ldots$. An algebra E over K is said to satisfy a polynomial identity if there exists $f(\zeta_1, \zeta_2, \ldots, \zeta_n) \in K[\zeta_1, \zeta_2, \ldots]$, $f \neq 0$ with $f(\alpha_1, \alpha_2, \ldots, \alpha_n) = 0$ for all $\alpha_1, \alpha_2, \ldots, \alpha_n \in E$. For example, any commutative algebra satisfies $f(\zeta_1, \zeta_2) = \zeta_1 \zeta_2 - \zeta_2 \zeta_1$. The first result on polynomial identities in group rings is fairly elementary.

THEOREM 5 (KAPLANSKY [5], AMITSUR [2]). *Let G have an abelian subgroup A with $[G:A] = n < \infty$. Then $K[G]$ satisfies a polynomial identity of degree $2n$*.

The converse direction appears to be much more difficult. We have

THEOREM 6 (ISAACS-PASSMAN [4]). *Let K have characteristic 0 and let $K[G]$*

satisfy a polynomial of degree n. Then G has an abelian subgroup A with $[G:A] \leq ([n/2]^2)!^{[n/2]}$.

The original proof of this result required the use of Jordan's theorem on finite complex linear groups and the bound on the index $[G:A]$ was in fact given by the function $J(n)$ associated with that theorem. In [13] an alternate argument was proposed and the bound given above is based on it.

Results in characteristic $p > 0$ are quite new. It was observed by Martha Smith that the Δ-techniques could be applied to this problem and she proved

Δ-THEOREM 7 (M. SMITH [14]). *Let $K[G]$ be a prime group ring satisfying a polynomial identity of degree n. Then Δ is torsion free abelian and $[G:\Delta] \leq n/2$.*

This result then motivated a more elementary combinatorial approach to the problem yielding the following two theorems. Here $\mathfrak{A}(n)$ is the integer valued function defined by

$$\mathfrak{A}(n) = (n!)^{n \cdot (n!)^2}.$$

Δ-THEOREM 8 (PASSMAN [12]). *Let $K[G]$ satisfy a polynomial identity of degree n. Then $[G:\Delta] \leq n!$.*

Δ-THEOREM 9 (PASSMAN [12], [13]). *Let $K[G]$ be semiprime and satisfy a polynomial identity of degree n. Then G has an abelian subgroup A with $[G:A] \leq n!\mathfrak{A}(n)$.*

If $K[G]$ is not semiprime, then the problem is still quite open. The following unpleasant example indicates the difficulties that arise. Let K have characteristic p and let P be an infinite extra-special p-group. Then $K[P]$ is Lie nilpotent of class p and hence satisfies a polynomial identity of degree $p + 1$. On the other hand, P has no abelian subgroup of finite index. Thus at present it is not even clear what the analogue of Theorem 9 should be if $K[G]$ is not semiprime. There are, however, some results of interest.

Let G be an arbitrary group and let $S(G)$ be the subgroup of G generated by all normal solvable subgroups of G. It follows easily that $S(G)$ is a characteristic locally solvable subgroup of G. In particular, if G is finite, then $S(G)$ is the unique maximal normal solvable subgroup of G.

Δ-THEOREM 10 (PASSMAN [13]). *Let $K[G]$ satisfy a polynomial identity of degree n. Then $[G:S(G)] \lneq n!\mathfrak{A}(n)$.*

It appears now that the next step must certainly be the study of finite p-groups in characteristic p and some partial results have already been obtained. For example, we have

PROPOSITION. *Let P be a finite nonabelian p-group and let K be any field. If $K[P]$ satisfies a polynomial identity of degree n then $n > (6p)^{1/3}$.*

Generalized polynomial identity rings. Let E be an algebra over K. A generalized polynomial over E is, roughly speaking, a polynomial in the indeterminates

$\zeta_1, \zeta_2, \ldots, \zeta_n$ in which elements of E are allowed to appear both as coefficients and between the indeterminates. We say E satisfies a generalized polynomial identity if there exists a nonzero generalized polynomial $f(\zeta_1, \zeta_2, \ldots, \zeta_n)$ such that $f(\alpha_1, \alpha_2, \ldots, \alpha_n) = 0$ for all $\alpha_1, \alpha_2, \ldots, \alpha_n \in E$. The problem here is precisely what does it mean for f to be nonzero. For example, suppose the center of E is bigger than K and let α be a central element not in K. Then E satisfies the identity $f(\zeta_1) = \alpha\zeta_1 - \zeta_1\alpha$ but surely this must be considered trivial. Again suppose that E is not prime. Then we can choose nonzero $\alpha, \beta \in E$ such that E satisfies the identity $f(\zeta_1) = \alpha\zeta_1\beta$ and this must also be considered trivial. We avoid these difficulties by restricting the allowable form of the polynomials.

We say that f is a weak polynomial of degree n if

$$f(\zeta_1, \zeta_2, \ldots, \zeta_n) = \sum_{\sigma \in S_n} \alpha_{1,\sigma} \zeta_{\sigma(1)} \alpha_{2,\sigma} \zeta_{\sigma(2)} \cdots \alpha_{n,\sigma} \zeta_{\sigma(n)} \alpha_{n+1,\sigma}$$

where $\alpha_{i,\sigma} \in E$. The above f is said to be nondegenerate if for some $\sigma, \alpha_{1,\sigma} E \alpha_{2,\sigma} E \cdots \alpha_{n,\sigma} E \alpha_{n+1,\sigma} \neq 0$. Otherwise f is degenerate.

Δ-THEOREM 11 (PASSMAN [13]). *Suppose that $K[G]$ satisfies a nondegenerate weak polynomial identity of degree n. Then $[G:\Delta] < \infty$ and $K[\Delta]$ satisfies a nondegenerate weak polynomial identity of degree n.*

Let E be an algebra over K. We say that E has a polynomial part if and only if E has an idempotent e such that eEe satisfies a polynomial identity.

Δ-THEOREM 12 (PASSMAN [13]). *Let $K[G]$ be a semiprime group ring. The following are equivalent:*
  (i) $[G:\Delta] < \infty$ *and* $|\Delta'| < \infty$.
  (ii) $K[G]$ *has a polynomial part.*
  (iii) $K[G]$ *satisfies a nondegenerate weak polynomial identity.*

Semisimple rings. A ring R is semisimple if its Jacobson radical JR is zero. The question of finding necessary and sufficient conditions for $K[G]$ to be semisimple is probably the most difficult of the problems I have mentioned. The best results to date are

THEOREM 13 (AMITSUR [1]). *Let K have characteristic 0 and suppose that K is not algebraic over the rationals Q. Then $K[G]$ is semisimple.*

THEOREM 14 (PASSMAN [10]). *Let K have characteristic $p > 0$ and suppose that K is not algebraic over $GF(p)$. If G has no elements of order p then $K[G]$ is semisimple.*

It is conjectured that the Jacobson radical of a finitely generated algebra is always a nil ideal. If this is proved, then the assumptions above that K is not algebraic over the prime field can be removed. The obvious converse to Theorem 14 is decidedly false. Let Z denote the infinite cyclic group and let Z_p be cyclic of order p. Then it is shown in [7] that $K[Z \wr Z_p]$ and $K[Z_p \wr Z]$ are both semisimple for any field K.

There are a number of results in the literature which guarantee that $JK[G] = 0$ under suitable hypotheses. But these do not help in formulating a conjecture as to the precise necessary and sufficient conditions required for semisimplicity. On the other hand, one result which might be suggestive is

THEOREM 15 (PASSMAN [8], [11]). *Let K have characteristic $p > 0$ and let G be metabelian. Then $JK[G] \neq 0$ if and only if G has a subgroup H satisfying*
  (i) *H is finite and $p\||H|$.*
  (ii) *$N = N(H)$ is normal in G.*
  (iii) *G/N is locally finite.*

ADDED IN PROOF. The author has recently obtained necessary and sufficient conditions for $K[G]$ to satisfy a polynomial identity. The result will appear in the Journal of Algebra.

REFERENCES

1. S. A. Amitsur, *On the semi-simplicity of group algebras*, Michigan Math. J. **6** (1959), 251–253. MR **21** #7256.

2. ———, *Groups with representations of bounded degree*. II, Illinois J. Math. **5** (1961), 198–205. MR **23** #A225.

3. I. G. Connell, *On the group ring*, Canad. J. Math. **15** (1963), 650–685. MR **27** #3666.

4. I. M. Isaacs and D. S. Passman, *A characterization of groups in terms of the degrees of their characters*, Pacific J. Math. **15** (1965), 877–903. MR **33** #199.

5. I. Kaplansky, *Groups with representations of bounded degree*, Canad. J. Math. **1** (1949), 105–112. MR **10**, 428.

6. D. S. Passman, *Nil ideals in group rings*, Michigan Math. J. **9** (1962), 375–384. MR **26** #2470.

7. ———, *On the semisimplicity of modular group algebras*, Proc. Amer. Math. Soc. **20** (1969), 515–519. MR **38** #4576.

8. ———, *On the semisimplicity of modular group algebras*. II, Canad. J. Math. **21** (1969), 1137–1145. MR **40** #1497.

9. ———, *Radicals of twisted group rings*, Proc. London Math. Soc. **20** (1970), 409–437.

10. ———, *On the semisimplicity of twisted group algebras*, Proc. Amer. Math. Soc. **25** (1970), 161–166.

11. ———, *Radicals of twisted group rings*. II, Proc. London Math. Soc. (to appear).

12. ———, *Linear identities in group rings*, Pacific J. Math. (to appear).

13. ———, *Linear identities in group rings*. II, Pacific J. Math. (to appear).

14. M. Smith, Ph.D. Thesis, University of Chicago, Chicago, Ill., 1970.

INSTITUTE FOR DEFENSE ANALYSES AND
UNIVERSITY OF WISCONSIN

Characters of Finite Groups and Sets of Primes

William F. Reynolds[1]

Let G be a finite group and π a set of primes. Feit [4] attempted to construct a representation theory for G with respect to π, but he ran into difficulties because of some questions about integral representations which are still unanswered as far as I know. I shall discuss a theory which avoids such questions by working entirely in terms of characters and class functions at the cost of weakening the results. Most of the results will be recognized as analogues of results of Brauer in the modular case $\pi = \{p\}$, and are proved by adapting some of his arguments. The general approach and many of the results presented here were worked out by Feit and by Brauer himself a long time ago.

Let $\mathscr{C}_G$ be the ring of (complex-valued) class functions on G; this is an inner-product space over $\mathbf{C}$ of dimension k, the class number of G. The set $\{\chi_i : 1 \leq i \leq k\}$ of irreducible characters is an orthonormal basis of $\mathscr{C}_G$ as well as a $\mathbf{Z}$-basis of the character ring $\mathscr{I}_G$. For the set $G_{\pi'}$ of all π'-elements of G, where π' is the set of primes not in π, define

$$\mathscr{C}_G(G_{\pi'}) = \{\alpha \in \mathscr{C}_G : \alpha(x) = 0 \text{ if } x \notin G_{\pi'}\},$$
$$P_G^\pi = \{\alpha \in \mathscr{I}_G : \alpha(x) = 0 \text{ if } x \notin G_{\pi'}\}.$$

P_G^π is a ring (without identity) with a $\mathbf{Z}$-basis $\{\Phi_j : 1 \leq j \leq l\}$, l being the number of conjugate classes in $G_{\pi'}$. The basis $\{\Phi_j\}$, like many things below, is determined only up to unimodular equivalence over $\mathbf{Z}$; this is one of the main weaknesses of the theory. In the modular case Brauer [2, Theorem 17] showed that the principal (i.e. projective) indecomposable characters form a canonical choice for $\{\Phi_j\}$. For arbitrary π, P_G^π is mentioned in Theorem 4 of Swan [9].

AMS 1970 *subject classifications*. Primary 20C20; Secondary 20C10.

[1] This work has been supported in part by the National Science Foundation.

$\{\Phi_j\}$ is a C-basis of $\mathscr{C}_G(G_{\pi'})$; let $\{\varphi_j\}$ be the dual basis defined by $(\varphi_j, \Phi_h) = \delta_{jh}$, and let M_G^π be the Z-module with basis $\{\varphi_j\}$. In the modular case the φ_j's, on $G_{p'}$, are just the irreducible Brauer characters [3, I, §3]. For $\pi = \emptyset$ we can take $\chi_i = \Phi_i = \varphi_i$.

$\Phi_j = \sum_i d_{ij}\chi_i$ where the decomposition matrix $D^\pi = (d_{ij})$ is a k by l matrix over Z with determinant divisors all 1. By orthogonality, $\chi_i(x) = \sum_j d_{ij}\varphi_j(x)$ if $x \in G_{\pi'}$. It follows that for $\beta \in \mathscr{C}_G(G_{\pi'})$, $\beta \in M_G^\pi$ if and only if there exists $\alpha \in \mathscr{I}_G$ such that $\beta(x) = \alpha(x)$ for all $x \in G_{\pi'}$. By this description (cf. [2, Theorem 15]) M_G^π is a ring with identity, and the l by l matrix $\boldsymbol{\varphi}^\pi$ of values of the φ_j's (cf. the table of irreducible Brauer characters) has entries in the ring $R(\pi')$ of algebraic integers in the field of $e_{\pi'}$th roots of unity, where $e_{\pi'}$ is the π'-part of the exponent of G.

If $x_1 = 1, x_2, x_3, \ldots$ are representatives of the conjugate classes of π-elements of G, then

$$\chi_i(x_m y) = \sum_j d_{ij}^m \varphi_j^m(y), \quad y \text{ any } \pi'\text{-element of } C_m = C_G(x_m),$$

where $\{\varphi_j^m\}$ is a Z-basis of $M_{C_m}^\pi$. This factorizes the character-table matrix χ as a product $\chi = D^\pi \boldsymbol{\varphi}^\pi$ of two k by k matrices, where $D^\pi = (d_{ij}^m)$ is the generalized decomposition matrix (with columns indexed by pairs (m,j)) and $\boldsymbol{\varphi}^\pi$ is the direct sum of the analogues of $\boldsymbol{\varphi}^\pi$ for the groups C_m (cf. [3, II, p. 45]). If $\tau \subseteq \pi$, set $\boldsymbol{\varphi}^\tau = D^{\tau,\pi}\boldsymbol{\varphi}^\pi$. For τ empty, $\boldsymbol{\varphi}^\tau = \chi$ and $D^{\tau,\pi} = D^\pi$, while for the set Π of all primes, we can take $\boldsymbol{\varphi}^\Pi$ as the identity matrix so that $D^{\pi,\Pi} = \boldsymbol{\varphi}^\pi$. Then $D^{\tau,\pi}$ always has entries in $R(\pi - \tau)$ [defined analogously to $R(\pi')$], and every prime ideal divisor of its determinant divides some prime in $\pi - \tau$. In particular we obtain information on D^π and $\boldsymbol{\varphi}^\pi$ (cf. [3, II, p. 45] and [3, I, (3E)]).

There are analogues of Brauer's theorem on characterization of characters for the rings P_G^π and M_G^π (of Z-rank l), as well as for certain larger rings $\mathscr{P}_G^\pi$ and $\mathscr{M}_G^\pi$ of Z-rank k whose bases consist of π-analogues of a generalization of Brauer characters [3, II, (7D)], [8]; for P_G^p this result is due to Brauer [1, Theorem 5]. These rings give rise to G-functors in the sense of Green [6].

In conclusion I refer briefly to the theory of π-blocks, which was first developed by Iizuka [7]. A π-block can be defined as a nonempty set of ordinary irreducible characters which is minimal for the property of being a union of p-blocks for every $p \in \pi$; there are a number of equivalent definitions. It is possible to assign φ_j's to π-blocks, etc. There is a π-block analogue of Brauer's second main theorem on blocks, implicit in [7]. Part of the theory of blocks and normal subgroups can be imitated for π-blocks, including analogues of Theorems 2B and 2D of Fong [5] (excluding the statements on defect groups). Finally, G has a normal π-complement if and only if the principal π-block contains just one φ_j.

References

1. R. Brauer, *Applications of induced characters*, Amer. J. Math. **69** (1947), 709–716. MR **9**, 268.
2. ———, *A characterization of the characters of groups of finite order*, Ann. of Math. (2) **57** (1953), 357–377. MR **14**, 844.
3. ———, *Zur Darstellungstheorie der Gruppen endlicher Ordnung*. I, Math. Z. **63** (1956), 406–444; ibid. II, Math. Z. **72** (1959/60), 25–46. MR **17**, 824; MR **21** #7258.
4. W. Feit, *Integral representations of finite groups*, Bull. Amer. Math. Soc. **63** (1957), 96. Abstract #164.
5. P. Fong, *On the characters of p-solvable groups*, Trans. Amer. Math. Soc. **98** (1961), 263–284. MR **22** # 11052.
6. J. A. Green, *Axiomatic representation theory for finite groups*, J. Pure Appl. Algebra **1** (1971), 41–77.
7. K. Iizuka, *Some studies on the orthogonality relations for group characters*, Kumamoto J. Sci. Ser. A **5** (1961), 111–118. MR **30** #2086.
8. W. F. Reynolds, *A generalization of Brauer characters*, Trans. Amer. Math. Soc. **119** (1965), 333–351. MR **31** #5899.
9. R. G. Swan, *The Grothendieck ring of a finite group*, Topology **2** (1963), 85–110. MR **27** #3683.

TUFTS UNIVERSITY

Bass-Orders and the Number of Nonisomorphic Indecomposable Lattices Over Orders

Klaus W. Roggenkamp

1. **Development of commutative Gorenstein-rings and commutative Bass-rings.**
In 1952 D. Gorenstein [13] considered—in a setting from algebraic geometry—the following class of commutative reduced noetherian rings Λ ("reduced" means without nilpotent elements): The integral closure Γ of Λ in its full ring of quotients A is finitely generated over Λ, and if $C = (\Lambda:\Gamma) = \{a \in A : a\Gamma \subset \Lambda\}$ is the conductor of Γ in Λ, then the length of Γ/Λ is equal to the length of Λ/C. This class of rings was called Gorenstein-rings by A. Grothendieck [14].

Independently H. Bass [3] in 1962 studied commutative noetherian rings Λ with finite self-injective dimension; i.e., $\text{inj.dim}_\Lambda({}_\Lambda\Lambda) < \infty$, where ${}_\Lambda\Lambda$ is the regular left Λ-module. J. P. Serre pointed out that these are Gorenstein-rings if they are reduced and if the integral closure in the total ring of quotients is finitely generated over Λ. In 1963, H. Bass [4] treated commutative Gorenstein-rings systematically. He gave seven equivalent definitions for Gorenstein-rings, of which we shall list two:

DEFINITION 1. A commutative noetherian ring Λ is called a *Gorenstein-ring*, if one of the following equivalent conditions is satisfied:

(i) For every $p \in \text{spec } \Lambda$, the prime spectrum of Λ, we have $\text{inj.dim}_{\Lambda_p}(\Lambda_p) < \infty$, where Λ_p is the localization of Λ at p.

(ii) For every $M \in {}_\Lambda M^f$, the category of finitely generated Λ-modules, grade $(\text{Ext}^j_\Lambda(M,\Lambda) \leq j$, for every $j \geq 0$. For $N \in {}_\Lambda M^f$, $\text{grade}(N) \geq i$ if $\text{Ext}^j_\Lambda(N,\Lambda) = 0$ for every $j < i$; i.e., $\text{grade}(N)$ is the least r for which $\text{Ext}^r_\Lambda(N,\Lambda) \neq 0$ (cf. Rees [19]).

If $\dim \Lambda = n < \infty$, where $\dim \Lambda$ denotes the Krull-dimension of Λ, then these two definitions are equivalent to $\text{inj.dim}_\Lambda({}_\Lambda\Lambda) < \infty$. (The Krull-dimension of a

---

AMS 1970 *subject classifications.* Primary 13H10, 13C10; Secondary 16A46, 16A16.

commutative noetherian ring Λ can be defined as the topological dimension of spec Λ, the prime spectrum, under the Zariski-topology.) This condition automatically implies inj.dim$_\Lambda(_\Lambda\Lambda) = n$.

Detour to noncommutative Gorenstein-rings. Auslander-Bridger [2], have recently defined noncommutative Gorenstein-rings. One can not use (i) in the noncommutative case, since there are no prime ideals and no localizations defined. (ii) does not involve concepts which can only be defined in the commutative case; however, in the noncommutative case, many of the nice properties of the "commutative" grade do not hold any longer; e.g., Λ a commutative ring, $M \in {}_\Lambda M^f$ with grade$(M) \geq i$ and $0 \to M' \to M \to M'' \to 0$ an exact sequence, implies grade(M'), grade$(M'') \geq i$. This fails in the noncommutative case. However, similar properties are essential in the theory of commutative Gorenstein-rings. It turns out that the proper generalization of commutative Gorenstein-rings is as follows (cf. also Auslander [1]): For Λ, a right noetherian ring, we say that $M \in {}_\Lambda M^f$ is *k-torsion*, ($k \in N$), if for every $M' \subset M$, we have grade $M' \geq k$.

DEFINITION 1'. A right noetherian ring Λ is called *left Gorenstein*, if for every $M \in {}_\Lambda M^f$, Ext$^i_\Lambda(M, {}_\Lambda\Lambda)$ is *i*-torsion for every $i \in N$. If Λ is left and right noetherian, then Λ is left Gorenstein if and only if it is right Gorenstein.

End of detour.

We return now to Bass' results on commutative Gorenstein-rings.

THEOREM 1 (BASS [4]). *Assume that Λ is a Gorenstein-ring (in the sense of Definition 1), satisfying:*

(a) dim $\Lambda \leq 1$.

(b) Λ *is reduced*.

(c) *The integral closure Γ of Λ in its total field of quotients is finitely generated over Λ.*

Then the following conditions are equivalent:

(i) *Every torsionless indecomposable Λ-module is isomorphic to an ideal.* (We recall that a Λ-module is called *torsionless*, if the natural map $M \to M^{**}$ is a monomorphism, where $M^* = $ Hom$_\Lambda(M, \Lambda)$.)

(ii) *If Λ' is a ring with $\Lambda \subset \Lambda' \subset \Gamma$, then Λ' is a Gorenstein ring.*

(iii) *Every Λ-ideal can be generated by two elements.*

EXAMPLES AND REMARK. If G is a finite abelian group, then $\Lambda = ZG$ is noetherian and dim $\Lambda = 1$, and inj.dim$_\Lambda(_\Lambda\Lambda) = 1$ (cf. later!). Thus Λ is a Gorenstein-ring. If Λ satisfies one of the equivalent conditions in the theorem, then $n(\Lambda) < \infty$, where $n(\Lambda)$ denotes the number of nonisomorphic indecomposable Λ-lattices. (A Λ-lattice is a finitely generated Λ-module which is Z-torsionfree.) The converse of this statement is not true, since for a *p*-group G, $p \neq 2$, $n(\Lambda) < \infty$ if and only if $|G| \leq p^2$, whereas Λ satisfies (i) if and only if $|G| = p$. Theorem 1 suggests a connection between being Gorenstein and the condition that all indecomposable torsionless modules are "small."

DEFINITION 2. Let Λ be a Gorenstein-ring, satisfying (a), (b), (c) of Theorem 1. If Λ has one of the equivalent conditions (i), (ii), or (iii), then Λ is called a *Bass-ring*.

This terminology is due to Drozd-Roĭter [**12**].

2. Bass-orders in separable finite-dimensional algebras.

NOTATION. K is a global field; i.e., K is either an algebraic number field or it is a finite field extension of the field of rational functions over a finite field, A is a finite-dimensional separable K-algebra; R is a Dedekind domain with quotient field K and Λ an R-order in A. We recall that an R-order Λ in A is a subring of A which contains a K-basis of A and which is an R-module of finite type. A Λ-lattice is a left Λ-module which is projective and of finite type over R.

In the theory of lattices over orders there was a different development, which led eventually to the concept of Bass-orders.

In 1965, Z. I. Borevich and D. K. Faddeev [**5**] generalized the following statement, which is a corollary of a paper by E. Dade, O. Taussky, and H. Zassenhaus [**8**]:

If $(A:K) = 2$, then every full Λ-ideal is invertible; whence projective. (If I is a full Λ-ideal, then $\Lambda(I) = \{a \in A : aI \subset I\}$ is an R-order in A containing Λ. I is said to be invertible if there exists a $\Lambda(I)$-ideal J such that $IJ = \Lambda(I)$.) This result implies in particular that every Λ-ideal has at most two generators. Then, if Γ is the maximal R-order in A, Γ/Λ is a cyclic Λ-module.

DEFINITION 3 (BOREVICH-FADDEEV [**5**]). Let A be a field. Then Λ is said to be of *cyclic index*, if Γ/Λ is a cyclic Λ-module, where Γ is the integral closure of R in A; i.e., the maximal R-order in A.

THEOREM 2 (BOREVICH-FADDEEV [**5**]). *Λ is of cyclic index if and only if every Λ-lattice is a direct sum of ideals.*

The *proof* uses essentially the fact that A is commutative; and in fact, it is not true in the noncommutative case. (For the noncommutative case, the definition must be changed appropriately!)

Looking back at Theorem 1, we see that Λ is of cyclic index if and only if Λ is a Bass-ring. Thus, here both developments meet, and A. V. Roĭter [**21**], Ju. A. Drozd, V. V. Kiričenko, and A. V. Roĭter [**11**] have pursued "Bass-orders" and have clarified the structure of Bass-orders locally.

DEFINITION 4. An R-order Λ in A is called a *Gorenstein-order* if one of the following equivalent conditions is satisfied:

(i) inj.dim$_\Lambda(_\Lambda\Lambda) = 1$ (cf. Definition 1, (i)),

(ii) $\Lambda^* = \text{Hom}_R(\Lambda_\Lambda, R)$ is a projective left Λ-module.

We call Λ a *Bass-order* if every R-order Λ' containing Λ is a Gorenstein-order (cf. Definition 2).

REMARKS. (i) inj.dim$_\Lambda(_\Lambda\Lambda) < \infty$ implies inj.dim$_\Lambda(_\Lambda\Lambda) = 1$.

(ii) Λ is "right" Gorenstein if and only if it is "left" Gorenstein.

EXAMPLES OF GORENSTEIN-RINGS ARE GROUP-RINGS. Let G be a finite group such that char $K \nmid |G|$; then $\Lambda = RG$ is an R-order in $A = KG$, which is Gorenstein. In fact, $\Lambda^* \cong \{a \in A : \text{Tr}_{A/K}(a\Lambda) \subset R\} = (1/|G|)\Lambda \cong \Lambda$. However, RG is usually not a Bass-order. As for commutative Bass-rings, one has

THEOREM 3 (ROĬTER [21]; DROZD-KIRIČENKO-ROĬTER [11]). *If Λ is a Bass-order in A, then every Λ-lattice is a direct sum of ideals.*

In connection with the work of Borevich-Faddeev and Dade-Taussky-Zassenhaus we have

THEOREM 4 (DROZD-KIRIČENKO-ROĬTER [11]). *An R-order Λ in A is a Bass-order if and only if the full two-sided Λ-ideals form a groupoid under proper multiplication.* (If I_1, I_2 are full two-sided Λ-ideals, then the product $I_1 I_2$ is proper, if for two-sided Λ-ideals $J_1 \supset I_1$ and $J_2 \supset I_2$ the equation $J_1 J_2 = I_1 I_2$ implies $I_1 = J_1$ and $I_2 = J_2$.)

Let us return to the result of Dade-Taussky-Zassenhaus [8], which inspired the work of Borevich-Faddeev.

THEOREM 5 (DADE-TAUSSKY-ZASSENHAUS [8]). *Let Λ be an R-order in the field A. If $(A:K) = n$, and if I is any nonzero Λ-ideal, then $I^{(n-1)}$ is invertible.*

REMARK AND PROBLEM. In view of Theorem 4, commutative Bass-orders are those where every ideal is invertible. In §4 we shall classify all those Bass-orders. From what we have seen up to now, a natural question arises: What are the Bass-orders of type n? Here an R-order Λ in the separable K-algebra A is a *Bass-order of type n*, if for every full two-sided Λ-ideal I, the nth power is invertible. Can this concept be used to give a definite, easy-to-handle answer to the problem of the finiteness of $n(\Lambda)$, the number of nonisomorphic indecomposable Λ-lattices?

3. Bass-orders and the number of indecomposable lattices.
We keep the notation of §2.

A still open problem is to classify those R-orders Λ in A for which $n(\Lambda) < \infty$. We have the following result of A. Jones [17], which localizes the problem. For $\boldsymbol{p} \in \operatorname{spec} R$, $\hat{X}_{\boldsymbol{p}}$ denotes the completion of $X \in M^f$ at $\boldsymbol{p}$.

LEMMA 1 (JONES [17]). *$n(\Lambda) < \infty$ if and only if $n(\hat{\Lambda}_{\boldsymbol{p}}) < \infty$ for all $\boldsymbol{p} \in \operatorname{spec} R$ for which $\hat{\Lambda}$ is not a Bass-order.*

We remark that this set of prime ideals is finite, and that for Bass-orders Λ, $n(\Lambda) < \infty$ by Theorem 3 and the Jordan-Zassenhaus theorem. (The Jordan-Zassenhaus theorem states that for an R-order Λ in the separable algebra A over a global field K, the set of Λ-lattices which span a fixed A-module, contains only finitely many nonisomorphic lattices.) For $n(\Lambda)$ we have the following partial result:

THEOREM 6 (DROZD-ROĬTER [12]; JACOBINSKI [16]). *If A is commutative then $n(\Lambda) < \infty$ if and only if $\mu_\Lambda(\Gamma/\Lambda) \leq 2$ and $\mu_\Lambda(\operatorname{rad}_\Lambda(\Gamma/\Lambda)) \leq 1$. Here $\mu_\Lambda(X)$ denotes the minimal number of generators of the Λ-module X, and $\operatorname{rad}_\Lambda(X)$ is the intersection of the maximal Λ-submodules of X.*

REMARK. The conditions given in Theorem 6 are those of Drozd-Roĭter. Jacobinski's conditions involve ramification indices, and they can be used to characterize those pairs (R,G), G a finite group, for which $n(RG) < \infty$. Assume that

R is the ring of algebraic integers in an algebraic number field. For every rational prime number p dividing the order of G, let $pR = \prod_j P_j^{e_j}$ be the prime decomposition of pR and put $e(p) = \max_j(e_j)$; and let G_p denote a p-Sylow-subgroup of G. Then $n(RG) < \infty$ if and only if for every $p\||G|$ either $e(p) = 1$ and G_p is cyclic of order p or p^2, or $e(p) \leq 2$ and G_p is cyclic of order p, or $p \leq 3$ and $e(p) = 3$ and G_p is cyclic of order p.

THE PROOF OF THEOREM 6 AND THE ROLE OF BASS-ORDERS. In view of Lemma 1 and since the conditions of Theorem 6 localize properly, it suffices to prove the theorem locally; i.e., we may assume that $\hat{R}$ is a complete discrete rank one valuation ring, the quotient field of which is a completion $\hat{K}$ of a global field. $\hat{\Lambda}$ is an $\hat{R}$-order in the separable finite-dimensional $\hat{K}$-algebra $\hat{A}$. Theorem 6 is then also valid, if one assumes that $\hat{A}$ is a direct sum of skewfields (cf. Roggenkamp [20]). However, the techniques employed in the proof do not carry over for arbitrary $\hat{A}$.

If we assume that one of the conditions of Theorem 6 is violated, then it is in principle easy, though laborious in practice, to construct an infinite sequence of nonisomorphic indecomposable $\hat{\Lambda}$-lattices, using a technique of Dade [7].

To prove the other direction, we may assume that $\hat{\Lambda}$ is a completely primary $\hat{R}$-order in $\hat{A}$. ($\hat{\Lambda}$ is called completely primary, if it is indecomposable as ring or equivalently, modulo its radical, it is a skewfield.) Moreover, the proof may be reduced to the case where $\hat{\Lambda}$ is not a Gorenstein-order. But then

$$\hat{\Lambda}_N = \{a \in \hat{A} : (\operatorname{rad} \hat{\Lambda}) a \subset \operatorname{rad} \hat{\Lambda}\}$$
$$= \{a \in \hat{A} : a \operatorname{rad} \hat{\Lambda} \subset \operatorname{rad} \hat{\Lambda}\}$$

is a Bass-order, if we assume $\mu_{\hat{\Lambda}}(\hat{\Gamma}/\hat{\Lambda}) \leq 2$ and $\mu_{\hat{\Lambda}}(\operatorname{rad}_{\hat{\Lambda}}(\hat{\Gamma}/\hat{\Lambda})) \leq 1$. We have to show that there are only finitely many nonisomorphic $\hat{\Lambda}$-lattices. We associate with every indecomposable $\hat{\Lambda}$-lattice $\hat{M}$ an exact sequence

$$E_{\hat{M}} : 0 \to \hat{M}' \to \hat{M} \to \hat{M}'' \to 0,$$

where $\hat{M}'' \cong \hat{X}^{(m)}$ and $\hat{X}$ is an indecomposable $\hat{\Lambda}$-module which depends only on $\hat{\Lambda}$ and not on $\hat{M}$. Moreover, $\hat{M}'$ is a lattice over a completely primary Bass-order $\hat{\Omega}$; e.g., if $\hat{A}$ is a skewfield, we associate with $\hat{M}$ the exact sequence

$$0 \to (\operatorname{rad} \hat{\Lambda})\hat{M} \to \hat{M} \to \hat{M}/(\operatorname{rad} \hat{\Lambda})\hat{M} \to 0.$$

Then $\hat{X} = \hat{\Lambda}/\operatorname{rad} \hat{\Lambda}$ and $\hat{\Omega} = \hat{\Lambda}_N$. If $\hat{A}$ is not a skewfield, one has to make slight modifications to ensure that $\hat{\Omega}$ is completely primary. To show that $n(\hat{\Lambda}) < \infty$ it suffices to establish that the degrees of the indecomposable $\hat{\Lambda}$-lattices are bounded. If one knows the indecomposable modules that can occur as summands of $\hat{M}'$ in the sequence $E_{\hat{M}}$, then the proof of Theorem 6 can be reduced to the decomposition of matrices, using a technique of Heller-Reiner [15].

Hence we have to classify completely primary Bass-orders $\hat{\Omega}$ and the indecomposable $\hat{\Omega}$-lattices. We recall that for a Bass-order $\hat{\Omega}$, $n(\hat{\Omega}) < \infty$ by Theorem 3.

4. **Classification of completely primary Bass-orders.** We assume that $\hat{\Lambda}$ is an $\hat{R}$-order in the separable finite-dimensional algebra $\hat{A}$ over a completion $\hat{K}$ of a global field.

THEOREM 7 (DROZD-KIRIČENKO-ROĬTER [11]). *Let $\hat{\Omega}$ be a Bass-order in $\hat{A}$, and let $\hat{M}$ be an indecomposable $\hat{\Omega}$-lattice. Then $\hat{M}$ is a projective $\hat{\Omega}'$-lattice for some $\hat{R}$-order $\hat{\Omega}'$ containing $\hat{\Omega}$. In addition, if $\hat{M}$ is faithful, then $\hat{M}'$ is a progenerator in the category of $\hat{\Omega}'$-lattices. ($\hat{M}'$ is called an $\hat{\Omega}'$-progenerator if $\mathrm{Hom}_{\hat{\Omega}'}(\hat{M}', -)$ is an exact and faithful functor. Two $\hat{R}$-order $\hat{\Omega}_1$ and $\hat{\Omega}_2$ are said to be Morita equivalent, if there exists an $\hat{\Omega}_1$-progenerator $\hat{E}$ such that $\mathrm{End}_{\hat{\Omega}_1}(\hat{E}) = \hat{\Omega}_2$. In this case there is a categorical equivalence between the left $\hat{\Omega}_1$-modules and the left $\hat{\Omega}_2$-modules, which preserves lattices etc.)*

Because of the Krull-Schmidt theorem, this result enables us to describe $n(\hat{\Omega})$ explicitly, once we know all R-orders $\hat{\Omega}'$ containing $\hat{\Omega}$. Before classifying the Bass-orders, let us state the analogue to Theorem 1 for Bass-orders.

THEOREM 8 (DROZD-KIRIČENKO-ROĬTER [11]). *Let $\hat{\Lambda}$ be a completely primary Bass-order in $\hat{A}$. Then the following conditions are equivalent:*

(i) *$\hat{\Lambda}$ is a Bass-order.*

(ii) *Every left $\hat{\Lambda}$-ideal has two generators.*

(iii) *Every $\hat{\Lambda}$-lattice is the direct sum of ideals.*

If $\hat{A} = \hat{D}$ or $\hat{A} = \hat{D}_1 \oplus \hat{D}_2$, where $\hat{D}, \hat{D}_1$ and $\hat{D}_2$ are complete skewfields, then the above conditions are equivalent to

(iv) *$\hat{\Gamma}/\Lambda$ is a cyclic $\hat{\Lambda}$-module, where $\hat{\Gamma}$ is the unique maximal $\hat{R}$-order in $\hat{A}$.*

If $\hat{A} = (\hat{D})_2$, where $\hat{D}$ is a complete skewfield, then the first three conditions are equivalent to

(v) *Every irreducible $\hat{\Lambda}$-lattice is cyclic.*

We *remark* that in view of the classification theorem (Theorem 9) $\hat{\Omega}$ can only be a completely primary Bass-order if $\hat{A} = \hat{D}_1$, $\hat{A} = \hat{D}_1 \oplus \hat{D}_2$ or $\hat{A} = (\hat{D}_1)_2$, where $\hat{D}_1$ and $\hat{D}_2$ are complete skewfields. Theorem 6 shows that $\hat{\Lambda}$ can not be too far off from being a Bass-order if $n(\hat{\Lambda}) < \infty$, and conversely.

THEOREM 9 (DROZD-KIRIČENKO-ROĬTER [11]); CLASSIFICATION THEOREM). *Let $\hat{\Omega}$ be a Bass-order in the separable finite-dimensional $\hat{K}$-algebra $\hat{A}$. Moreover, assume that $\hat{\Omega}$ is indecomposable as ring. Then $\hat{\Omega}$ must be of one of the following types:*

(I) *$\hat{\Omega}$ is Morita-equivalent to an $\hat{R}$-order $\hat{\Omega}_0$ in $\hat{D}_1 \oplus \hat{D}_2$, where $\hat{D}_1$ and $\hat{D}_2$ are complete skewfields and $\hat{\Omega}_0 e_1 = \hat{\Gamma}_i$, where $\hat{\Gamma}_i$ is the maximal $\hat{R}$-order in $\hat{D}_i$ and e_i is the idempotent corresponding to $\hat{D}_i$, $i = 1,2$.*

(II) *$\hat{\Omega}$ is Morita-equivalent to a Bass-order $\hat{\Omega}_0$ in a skewfield $\hat{D}$. Moreover, there exists a unique chain of over-orders*

$$\hat{\Omega}_0 \subsetneq \hat{\Omega}_1 \subsetneq \cdots \subsetneq \hat{\Omega}_s = \hat{\Gamma},$$

where $\hat{\Gamma}$ is the maximal $\hat{R}$-order in $\hat{D}$, and if $\hat{N} = \mathrm{rad}\,\Omega_0$, then $F_i = \hat{\Omega}_i/\hat{N}\hat{\Omega}_i$ is a two-dimensional $F = F_0$-vectorspace, which is also a ring; $F_i = F[r_i]$, $r_i^2 = 0$ and $Fr_i = r_iF$, $1 \leq i < s$. However, r does not necessarily lie in the center of F_i; the action of r on F is related to the Frobenius-automorphism of $\hat{D}$. For $F_s = \hat{\Gamma}/\hat{N\Gamma}$ we have to distinguish two cases:

(IIa) $F_s = F[r_s]$, $r_s^2 = 0$, $Fr_s = r_sF$,

(IIb) F_s is a two-dimensional extension field of F.

(III) $\hat{\Omega}$ is hereditary of type s. ($\hat{\Lambda}$ is called hereditary if every $\hat{\Lambda}$-lattice is projective and it is hereditary of type s if it has exactly s nonisomorphic indecomposable lattices or equivalently, $\hat{\Lambda}$ is contained in s different maximal orders.)

(IV) $\hat{\Omega}$ is Morita-equivalent to a completely primary Bass-order $\hat{\Omega}_0$ in $(\hat{D})_2$, $\hat{D}$ a skewfield; moreover, there exists a unique chain of over-orders

$$\hat{\Omega}_0 \subsetneqq \hat{\Omega}_1 \subsetneqq \cdots \subsetneqq \hat{\Omega}_s,$$

where $\hat{\Omega}_{s-1}$ is completely primary and $\hat{\Omega}_s$ decomposes as ring. Two cases can occur here:

(IVa) $\hat{\Omega}_s$ is maximal,

(IVb) $\hat{\Omega}_s$ is nonmaximal hereditary.

(V) $\hat{\Omega}$ is Morita-equivalent to the $\hat{R}$-order $\hat{\Omega}_0$ in $(\hat{D})_2$,

$$\hat{\Omega}_0 = \begin{bmatrix} \hat{\Gamma} & (\mathrm{rad}\ \hat{\Gamma})^d \\ \hat{\Gamma} & \hat{\Gamma} \end{bmatrix}, \quad d \geq 2,$$

where $\hat{\Gamma}$ is the maximal $\hat{R}$-order in $\hat{D}$.

REMARK. All these types of Bass-orders do exist if the proper conditions are satisfied; e.g., Bass-orders of type (IIb) exist if $\hat{\Gamma}/\mathrm{rad}\ \hat{\Gamma}$ contains a subfield of index 2. On the other hand all orders of the types (I)–(V) are Bass-orders.

Theorem 7 suggests that it might be possible to express the Grothendieck-group $G_0(\Lambda)$ of a Bass-order Λ in terms of the Grothendieck-groups of the maximal orders Γ lying over it. (The Grothendieck-group $G_0(\Lambda)$ of Λ-lattices relative to short exact sequences is the abelian group generated by symbols $[M]$, one for each Λ-lattice M and subject to the relations $[M] = [M'] + [M'']$ whenever there exists an exact sequence of Λ-lattices $0 \to M' \to M \to M'' \to 0$.) However, globally this can not be true as show the hereditary orders; but it is true locally except for Bass-orders of type (IIb). To be more precise, we have:

LEMMA 2. *Let $\hat{\Lambda}$ be a Bass-order in $\hat{A}$ which is indecomposable as ring. If $\{\hat{\Gamma}_i\}_{1 \leq i \leq n}$ are the different maximal $\hat{R}$-orders containing $\hat{\Lambda}$, then the map*

$$\varphi : \bigoplus_{i=1}^n G_0(\hat{\Gamma}_i) \to G_0(\hat{\Lambda}),$$

induced by restriction of the operators, is an epimorphism, except if $\hat{\Lambda}$ is of type (IIb).

Using this lemma and the classification theorem, one can compute $G_0(\hat{\Lambda})$ explicitly:

THEOREM 10. *Let $\hat{\Lambda}$ be a Bass-order which is indecomposable as ring. With the classification in Theorem 9 we have*:

(I) $G_0(\hat{\Lambda}) \cong Z^{(2)}$,
(IIa) $G_0(\hat{\Lambda}) \cong Z$,
(IIb) $G_0(\hat{\Lambda}) \cong Z \oplus Z/2Z$,
(III) $G_0(\hat{\Lambda}) \cong Z^{(s)}$,
(IVa) $G_0(\hat{\Lambda}) \cong Z$,
(IVb) $G_0(\hat{\Lambda}) \cong Z \oplus Z/2Z$,
(V) $G_0(\hat{\Lambda}) \cong Z^{(2)}$.

FINAL REMARK ON $n(\hat{\Lambda})$. Theorem 6 states that in the commutative case for $n(\hat{\Lambda})$ to be finite it is necessary and sufficient that $\hat{\Lambda}$ is not too far off from being Bass; but it need not be Bass. However, in the strictly noncommutative case, $\hat{A} = (\hat{K})_n$ Drozd-Kiričenko [10] have shown that in case $n = 2$, for an $\hat{R}$-order $\hat{\Lambda}$ in $\hat{A}$, $n(\hat{\Lambda}) < \infty$ if and only if $\hat{\Lambda}$ is a Bass-order. Moreover, in this case there is a largest $\hat{R}$-order $\hat{\Lambda}_0$ with $n(\hat{\Lambda}_0) = \infty$ but rad $\hat{\Lambda}_0$ is a lattice over a maximal order. On the other hand, in the commutative case the computations of Drozd [9] show that in case A is a three-dimensional semisimple Q-algebra, Λ an R-order in A, and Γ the maximal R-order in A, $n(\Lambda) < \infty$ if and only if $\mu_\Lambda(\text{rad}_\Lambda(\Gamma/\Lambda)) \leq 1$. However, in this case Λ is a Bass-order of type 2 (cf. Theorem 5). Apart from the fact that in the noncommutative case there are many maximal orders containing a given one, it does not seem probable that the conditions of Theorem 6 can be generalized; however, the methods employed in the proof can be generalized, and it seems probable that Bass-orders will also play a key role in the general situation.

REFERENCES

1. M. Auslander, *Anneaux de Gorenstein, et torsion en algèbre commutative*, Séminaire d'Algèbre Commutative dirigé par Pierre Samuel, 1966/67, École Normale Supérieure de Jeunes Filles, Secrétariat mathématique, Paris, 1967. MR **37** #1435.

2. M. Auslander and M. Bridger, *Stable module theory*, Mem. Amer. Math. Soc. No. 94 (1969).

3. H. Bass, *Injective dimension in Noetherian rings*, Trans. Amer. Math. Soc. **102** (1962), 18–29. MR **25** #2087.

4. ———, *On the ubiquity of Gorenstein rings*, Math. Z. **82** (1963), 8–28. MR **27** #3669.

5. Z. I. Borevič and D. K. Faddeev, *Representations of orders with a cyclic index*, Trudy Mat. Inst. Steklov. **80** (1965), 51–65 = Proc. Steklov Inst. Math. **80** (1965), 56–72. MR **34** #5805.

6. ———, *A remark on orders with a cyclic index*, Dokl. Akad. Nauk SSSR **164** (1965), 727–728 = Soviet Math. Dokl. **6** (1965), 1273–1274. MR **32** #7601.

7. E. C. Dade, *Some indecomposable group representations*, Ann. of Math. (2) **77** (1963), 406–412. MR **26** #2521.

8. E. C. Dade, O. Taussky, and H. Zassenhaus, *On the theory of orders, in particular on the semigroup of ideal classes and genera of an order in an algebraic number field*, Math. Ann. **148** (1962), 31–64. MR **25** #3962.

9. Ju. A. Drozd, *On the representations of cubic Z-rings*, Dokl. Akad. Nauk SSSR **174** (1967), 16–18 = Soviet Math. Dokl. **8** (1967), 572–574. MR **35** #6659.

10. Ju. A. Drozd and V. V. Kiričenko, *Representation of rings in a second order matrix algebra*, Ukrain. Mat. Ž. **19** (1967), No. 3, 107–112. (Russian) MR **35** #1632.

11. Ju. A. Drozd, V. V. Kiricenko, and A. V. Roĭter, *On hereditary and Bass orders*, Izv. Akad. Nauk SSSR Ser. Mat. **31** (1967), 1415–1436 = Math. USSR Izv. **1** (1967), 1357–1376. MR **36** #2608.

12. Ju. A. Drozd and A. V. Roĭter, *Commutative rings with a finite number of indecomposable integral representations*, Izv. Akad. Nauk SSSR Ser. Mat. **31** (1967), 783–798 = Math. USSR Izv. **1** (1967), 757–772. MR **36** #3768.

13. D. Gorenstein, *An arithmetic theory of adjoint plane curves*, Trans. Amer. Math. Soc. **72** (1952), 414–436. MR **14**, 198.

14. A. Grothendieck, *Théorèmes de dualité pour les faisceaux algébriques cohérents*, Séminaire Bourbaki 1957–1962, Secrétariat mathématique, Paris, 1962. MR **26** #3566.

15. A. Heller and I. Reiner, *Representations of cyclic groups in rings of integers*. I, II, Ann. of Math. (2) **76** (1962), 73–92; ibid. (2) **77** (1963), 318–328. MR **25** #3993; MR **26** #2520.

16. H. Jacobinski, *Sur les ordres commutatifs avec un nombre fini de réseaux indécomposables*, Acta Math. **118** (1967), 1–31. MR **35** #2876.

17. A. Jones, *Groups with a finite number of indecomposable integral representations*, Michigan J. Math. **10** (1963), 257–261. MR **27** #3698.

18. L. A. Nazarova and A. V. Roĭter, *A sharpening of a theorem of Bass*, Dokl. Akad. Nauk SSSR **176** (1967), 266–268 = Soviet Math. Dokl. **8** (1967), 1089–1092. MR **37** #1402.

19. D. Rees, *The grade of an ideal or module*, Proc. Cambridge Philos. Soc. **53** (1957), 28–42. MR **18**, 637.

20. K. W. Roggenkamp, *A necessary and sufficient condition for orders in direct sums of complete skewfields to have only finitely many nonisomorphic indecomposable representations*, Bull. Amer. Math. Soc. **76** (1970), 130–134.

21. A. V. Roĭter, *An analog of the theorem of Bass for modules of representations of noncommutative orders*, Dokl. Akad. Nauk SSSR **168** (1966), 1261–1264 = Soviet Math. Dokl. **7** (1966), 830–833. MR **34** #2632.

MCGILL UNIVERSITY, MONTREAL, AND
UNIVERSITY OF BIELEFELD, GERMANY

The Modular Theory of Permutation Representations

L. Scott

Before entering upon a detailed description of our theory,[0] we mention two of its consequences which can be stated in the classical language of permutation groups and modular representation theory.

Let G be a finite group acting on a finite set Ω (not necessarily transitively or faithfully), and let θ be the associated permutation character. For p a prime and B a p-block, set $\theta_B = \sum_{\chi \in B} (\theta, \chi) \chi$.

COROLLARY A. *Suppose $x \in G$ and a power x^n is a p-element. Let $\tilde{\theta}$ be the permutation character for the action of $C_G(x^n)$ on the set of fixed points of x^n in Ω. Then*

$$\theta_B(x) = \sum_{b^G = B} \tilde{\theta}_b(x).$$

In particular, $\theta_B(x^n)$ is a nonnegative rational integer, and $|\theta_B(x)| \leqq \theta_B(x^n) \leqq \theta(x^n)$.

It is not hard to prove Corollary A directly (using Nagao's proof of Brauer's second fundamental theorem[1]). The result below, however, does not seem to be a consequence of the classical theory.

COROLLARY B. *Suppose the block B has defect 1. Let $\langle x \rangle$ be a defect group of B, and let b be the block of $N_G(\langle x \rangle)$ satisfying $b^G = B$. Let $\tilde{\theta}$ be the permutation character of $N_G(\langle x \rangle)$ on the fixed points of x in Ω, and set $\tilde{\theta}_b = \sum_j l_j \tilde{\chi}_j$, where $\tilde{\chi}_j$ is absolutely irreducible and appears in $\tilde{\theta}_b$ with multiplicity l_j.*

Then we may write $\theta_B = \sum_j l_j \Psi_j + \Phi$ where

(1) Φ *is a sum (possibly zero) of projective indecomposable characters of B,*

(2) $\Psi_j(x) > 0$ *for each j, and Ψ_j is either a nonexceptional character of B or the sum of all exceptional characters of B.*

AMS 1970 *subject classifications.* Primary 20C20; Secondary 16A26, 20C05.

[0] The present work is to some extent foreshadowed by Scott [**1**, §5, §6, 14.].

[1] Also, Corollary A implies Brauer's second fundamental theorem, when applied to the natural action of $G \times G$ on G.

REMARK. In general $\Psi_j = \Psi_k$ with $j \neq k$ is possible (though Ψ_j is the trivial character if and only if $\tilde{\chi}_j$ is the trivial character). l_j is, in fact, a "multiplicity," but of an indecomposable module, not a character.

Corollary B is a consequence of a theorem in §6 describing the characters of indecomposable modules in a "relative" defect 1 situation.

Corollary A is also a consequence of a theorem on the characters of indecomposable modules (see §5).

0. **The centralizer ring.** If S is any commutative ring with identity, we define the S-centralizer ring $V_S(G) = V_S(G,\Omega)$ to be the collection of all matrices with entries from S that commute with the permutation matrices determined (with respect to some fixed ordering of Ω) by elements of G. In case S is the ring of rational integers, we write only $V(G)$ for $V_S(G)$ and refer to $V(G)$ as the *centralizer ring*.[2] The *standard basis matrices* $\{A_i\}_{i=1}^r$ are obtained from the full set $\{O_i\}_{i=1}^r$ of orbits of G on $\Omega \times \Omega$ by setting the α, β entry of A_i equal to 1 for $(\alpha,\beta) \in O_i$ and 0 otherwise. These matrices always form an S-basis for $V_S(G)$. In particular, $V_S(G)$ is isomorphic to the tensor product $SV(G)$.

The notation $S\Omega$ refers to the usual S-free SG module obtained from the action of G on Ω; we regard $S\Omega$ also as a $V_S(G)$ module in the obvious way.[3]

In this paper S will usually be K, R, or F where K is a p-adic number field (p a fixed prime), R is the ring of local integers in K, and F is the residue class field $R/\pi R$. π is a generator of the maximal ideal of R. We use the notation $\bar{x}$ for the image of x under some (hopefully obvious) map $X \to X/\pi X$ of an R-module X containing x.

The isomorphism mentioned in the first paragraph implies

$$\overline{V_R(G)} \cong V_F(G).$$

As a consequence, the RG-indecomposable components of $R\Omega$ display none of the pathologies possible (see Feit's notes; III, example following 3.17) for the general indecomposable RG module.

PROPOSITION 1. *Let M, N be RG-indecomposable components of $R\Omega$. Then*

(a) $\overline{\mathrm{Hom}_{RG}(M, N)} \cong \mathrm{Hom}_{FG}(\overline{M}, \overline{N})$.
(b) $\overline{M}$ *is indecomposable and has the same vertex as M.*

1. **Decomposition numbers.** By using the primitive idempotents in $V_R = V_R(G,\Omega)$ we can establish a 1-1 correspondence $[M_j] \leftrightarrow [U_j]$ between isomorphism classes of RG-indecomposable components of $R\Omega$ and isomorphism classes of projective

---

[2] See Wielandt [1, V] for the classical theory of permutation representations and their centralizer rings.

[3] All "modules" are, by convention, finitely generated and acted upon on the right, with the exception that the action of a base ring such as S is written on the left.

indecomposable V_R modules. Similarly there is a 1-1 correspondence $[X_s] \leftrightarrow [Z_s]$ between isomorphism classes of KG-irreducible submodules of $K\Omega$ and isomorphism classes of irreducible V_K modules. Finally, there is a well-known 1-1 correspondence $[U_j] \leftrightarrow [L_j]$ of isomorphism classes of projective indecomposable V_R modules and isomorphism classes of irreducible V_F modules.

In view of these correspondences, there are three kinds of "decomposition numbers" we can define:

(1) Let Z_s^0 be an R-free V_R module with $KZ_s^0 \approx Z_s$.
Then we list the composition factors of $\bar{Z}_s^0$, writing $\bar{Z}_s^0 \leftrightarrow \sum_j d_{sj} L_j$.
(2) Set $KU_j \approx \sum_s d_{sj}^V Z_s$.
(3) Set $KM_j \approx \sum_s d_{sj}^\Omega X_s$.

THEOREM 1. *Assume that K,F are splitting fields for V_K, V_F respectively. Then $d_{sj} = d_{sj}^V = d_{sj}^\Omega$. Furthermore $R\Omega \approx \sum_j (\dim L_j) M_j$.*

This purely formal result depends only on the complete reducibility of $K\Omega$. The first equality is well known (see IX.8 in Artin-Nesbitt-Thrall).

2. **Defect groups.** Let A_i be a standard basis matrix. We define "the" *defect group*[4] D_i of A_i to be a sylow p-subgroup of $G_{\alpha\beta}$ where $(\alpha,\beta) \in U_i$. (Defect groups are well defined only up to conjugation by elements of G.)

For any p-subgroup P of G we define $I_F(P) = I_F(P; G,\Omega)$ to be the set of F-linear combinations of A_i's satisfying $D_i \leq_G P$. Then the following classical[5] lemma holds.

LEMMA 1. *For P, Q p-subgroups of G*

$$I_F(P) I_F(Q) \leq \sum_{\substack{D_i \leq_G P \\ D_i \leq_G Q}} I_F(D_i).$$

Now, given a primitive idempotent $e \in V_F$, there is an index i such that $e \in I_F(D_i)$ and $D_i \leq_G P$ whenever $e \in I_F(P)$. We set $D(e) =_G D_i$ and call $D(e)$ "the" *defect group of e*. The ideals $I_F(P)$ are easily found to be "relative trace ideals" (see Green [3]) and so Green's work yields immediately the following proposition.

PROPOSITION 2. *$D(e)$ is the vertex of $F\Omega e$.*

If F is a splitting field for V_F and $e \in V_F$ is a primitive idempotent, then there is a unique irreducible modular character λ of V_F satisfying $\lambda(e) = 1$. This leads to a second characterization[6] of defect groups.

---

[4]This definition remedies here the difficulty mentioned by Green [3, footnote 2]. The D_i's directly generalize the concept of "conjugacy class defect group."

[5]See Osima [1, Lemma 4], which is based on Brauer [3, p. 112]. Green [3, 4.11] is very enlightening.

[6]The classical versions are Osima [1, Corollary 2], Rosenberg [1, Proposition 3.2], and, originally, Brauer [3, Theorem 3].

PROPOSITION 3. *Assume F is a splitting field for V_F. Let $e, e' \in V_F$ be primitive idempotents, and let λ, λ' be the associated modular characters. Then*
(1) $\lambda(xA_i) = 0$ *for all $x \in V_F$ unless $D_i \geqq_G D(e)$.*
(2) $\lambda(A_i) \neq 0$ *for some i with $D_i =_G D(e)$.*
(3) *e is equivalent to e' if and only if $D(e) =_G D(e')$ and $\lambda(A_i) = \lambda'(A_i)$ for all i with $D_i =_G D(e)$.*

The following theorem is slightly out of context in that its proof depends on Theorem 3 in §4.

THEOREM 2. *A necessary condition that D be the vertex of an indecomposable component of $F\Omega$ is that there exist $\alpha \in \Omega$ and $y \in N_G(D)$ such that D is a sylow p-subgroup of $G_{\alpha\alpha y}$.*

3. The Brauer homomorphism. Suppose $P \leqq H \leqq N_G(P)$ where P is a p-group. Then there is a natural homomorphism $f = f_{(G,\Omega,P,H)}$ mapping $V_F(G,\Omega)$ into $V_F(H,\Omega_P)$:[7] First define f in the special case $G = H$—if $O_i \subseteq \Omega_P \times \Omega_P$ define $f(A_i)$ to be the standard basis matrix for (H,Ω_P) corresponding to O_i, and otherwise set $f(A_i) = O$; then extend the definition of f to $V_F(H,\Omega)$ by linearity. f is now defined in the general case as the composite of the natural restriction map $V_F(G,\Omega) \to V_F(H,\Omega)$ and $f_{(H,\Omega,P,H)}$.

The relationship between f and the classical Brauer homomorphism[8] is described below.

PROPOSITION 4. *Assume $C_G(P) \leqq H$ and let $s: Z(FG) \to Z(FH)$ be the standard Brauer homomorphism. Then the following diagram is commutative:*

$$\begin{array}{ccc} Z(FG) & \to & V_F(G,\Omega) \\ s \downarrow & & f \downarrow \\ Z(FH) & \to & V_F(H,\Omega_P) \end{array}$$

The following two propositions describe elementary properties of $f_{(G,\Omega,P,H)}$.[9]

PROPOSITION 5. *Let $e \in V_F(G)$ be a primitive idempotent. Then $f(e) \neq 0$ if and only if $D(e) \geqq_G P$.*

PROPOSITION 6. *Let $e \in V_F(G)$ be a primitive idempotent with $f(e) \neq 0$. If N is an indecomposable component of $F\Omega_P f(e)$, then some $D(e)$ contains a vertex of N. If $C_G(P) \leqq H$ then N lies in a block b of H such that $F\Omega e$ is in b^G.*[10]

---

[7]Ω_P denotes the set of fixed points of P in Ω.

[8]Very good insight into the connection between our theory and the classical theory is obtained by considering the action of $G \times G$ on G. (This point of view originates with Green [1], [3].) $V_F(G \times G, G) \cong Z(FG)$, standard basis matrices correspond to class sums, defect groups correspond, and the homomorphism f becomes the standard Brauer homomorphism.

[9]The approach here is modeled on Rosenberg [1] and Feit [1].

[10]See Brauer [6, §2] for a definition of b^G.

It is also worth noting the obvious fact that any indecomposable component of $F\Omega_P$ is isomorphic to a component of $F\Omega_P f(e)$ for some primitive $e \in V_F(G)$.

At this stage of our development we sacrifice a little simplicity for the sake of obtaining more detailed information.

The 1-1 condition. A p-group $Q \geq_G P$ satisfies the 1-1 condition with respect to (G,Ω,P,H) if $P \leq_G D_i$ whenever $D_i \leq_G Q$.

The onto condition. A p-group $Q \geq_G P$ satisfies the onto condition with respect to (G,Ω,P,H) if $o^G \cap \Omega_P \times \Omega_P = o$ whenever o is an orbit of H on $\Omega_P \times \Omega_P$ whose associated defect group d satisfies $d \leq_G Q$.

If P contains no proper subgroup conjugate to any D_i, then P satisfies the 1-1 condition with respect to (G,Ω,P,H).

If $H = N_G(P)$, and if $P = Q$ (or more generally, if P is weakly closed in Q with respect to G), then Q satisfies the onto condition with respect to (G,Ω,P,H).

If e is any idempotent in $V_F(G)$, we let Ψ_e denote the character of G afforded by $K\Omega e_0$ where e_0 is an idempotent in $V_R(G)$ satisfying $\bar{e}_0 = e$. In addition we define $\Psi_e = 0$ for $e = 0$.

PROPOSITION 7. *Suppose Q satisfies the 1-1 condition with respect to (G,Ω,P,H). Let e,e' be idempotents in $V_F(G)$, and assume that e is primitive with $D(e) =_G Q$. Then*

$$(\Psi_e, \Psi_{e'})_G \leq (\Psi_{f(e)}, \Psi_{f(e')})_H.$$

PROPOSITION 8. *Suppose Q satisfies the onto condition with respect to (G,Ω,P,H). Let e, e' be idempotents in $V_F(G)$, and assume that e is primitive with $D(e) =_G Q$. Then*

$$(\Psi_e, \Psi_{e'})_G \geq (\Psi_{f(e)}, \Psi_{f(e')})_H.$$

The next section is devoted to further consequences of the onto condition.

4. **Brauer's first fundamental theorem.**[11] We *assume throughout this section* that Q satisfies the onto condition with respect to (G,Ω,P,H). $P \leq H \leq N_G(P)$ and $f = f_{(G,\Omega,P,H)}$ as in §3.

THEOREM 3. (a) *Let e, e' be idempotents in $V_F(G,\Omega)$ with e primitive and $D(e) =_G Q$. Then $f(e)$ is primitive, and $F\Omega e | F\Omega e'$ if and only if $F\Omega_P f(e) | F\Omega_P f(e')$.*

(b) *In case $H = N_G(P)$ and $\tilde{e} \in V_F(H, \Omega_P)$ is a primitive idempotent with $D(\tilde{e}) = P$, then $\tilde{e}$ is equivalent to $f(e)$ for some primitive $e \in V_F(G)$ with $D(e) =_G P$.*

The following result depends on a lemma in the next section.

PROPOSITION 9. *Suppose $H = N_G(P)$ and $e \in V_F(G)$ is a primitive idempotent with $D(e) =_G P$. Then $F\Omega_P f(e)$ is the Green correspondent*[12] *of $F\Omega e$. In particular*

---

[11]See Brauer [5], and Rosenberg [1].
[12]Green [2, Theorem 2].

$$\Psi_e(1) \equiv [G:H]\Psi_{f(e)}(1) \qquad (p^{v_p([G:P])+1}).$$

The original motivation for the present theory was the desire to (1) simplify the problem of computing the (ordinary) character table of $V(G)$, and (2) obtain information on primes dividing the order of G from knowledge of the character table of $V(G)$. In these respects the result below is fundamental. Generous assumptions have been made on F to avoid uninteresting complications.

THEOREM 4. *Assume F is a splitting field for $V_F(G,\Omega)$ and $V_F(H,\Omega_P)$. Suppose $e \in V_F(G)$ is a primitive idempotent with $D(e) =_G Q$ and let λ, $\tilde{\lambda}$ be the modular characters[13] associated with $e, f(e)$ respectively. Then $\lambda = \tilde{\lambda} \circ f$.*

5. **Brauer's second fundamental theorem.**[14] Again, $P \leq H \leq N_G(P)$ where P is a p-subgroup of G, and f denotes $f_{(G,\Omega,P,H)}$.

LEMMA 2. *Let $e \in V_F(G)$ be an idempotent. Then $(F\Omega e)_H \approx F\Omega_P f(e) \oplus T$ where T is a summand of $F(\Omega - \Omega_P)$.*

THEOREM 5. *Suppose $x \in G$ and a power $x^n = z$ is a p-element. Let $e \in V_F(G)$ be an idempotent, and take $P = \langle z \rangle$, $x \in H$. Then $\Psi_e(x) = \Psi_{f(e)}(x)$. In particular $\Psi_e(z)$ is a nonnegative rational integer· and $|\Psi_e(x)| \leq \Psi_e(z) \leq \theta(z)$.*

If e is primitive, then $\Psi_e(z) > 0$ if and only if z is conjugate in G to an element of $D(e)$.

The reader will note that Corollary A is a consequence of the above theorem and Proposition 6 in §3.

6. **Defect 0 and 1.**[15] Assume K, F are splitting fields for V_K, V_F, and let χ_s, Ψ_j denote the characters of G afforded by X_s, KM_j respectively (see §1 for notation). Let the symbol $\sum_p \chi_s^\sigma$ denote the sum of all distinct p-conjugates of χ_s, and write $s \sim t$ if χ_s is p-conjugate to χ_t.

Set $a = v_p(|G|)$, and define $v_s = v_p(\chi_s(1))$. Put $e = \max\{v_p(|O_i|)\}_{i=1}^r$.

By a theorem of Wielandt (see Keller [1]), we have $v_s \leq e$ for all s. We are interested here in the cases $v_s = e$ and $v_s = e - 1$.

LEMMA 3. *Suppose the number of p-conjugates of χ_s is divisible by p^y. Then*
(a) $y \leq e - v_s$.
(b) *If $y = e - v_s$ then χ_s has exactly p^y p-conjugates, and $\sum_p \chi^\sigma = \Psi_j$ for some j. If $\chi_s \subseteq \Psi_k$ then $k = j$. The vertex of M_j has order p^{a-e}.*

THEOREM 6. *Suppose $v_s = e$.*
Then $\chi_s = \Psi_j$ for some j. If $\chi_s \subseteq \Psi_k$ then $k = j$. The vertex of M_j has order p^{a-e}. Also, χ_s is p-rational.

---

[13] Note that these characters do not always have degree 1, which is the classical case (see Brauer [5, 7D]).

[14] See Brauer [6]. The approach here is modeled on Nagao [1].

[15] This section is analogous in spirit and content to Brauer [1].

THEOREM 7. *Suppose* $v_s = e - 1$. *Then we have either* A *or* B *below*:

(A) χ_s *has exactly* p *p-conjugates.* $\sum_p \chi_s^\sigma = \Psi_j$ *for some* j, *and if* $\chi_s \subseteq \Psi_k$, *then* $k = j$. *The vertex of* M_j *has order* p^{a-e}.

(B) *The number of p-conjugates of* χ_s *divides* $p - 1$. *If* $\chi_s \subseteq \Psi_j$ *then one of the following occurs*:

(i) $\Psi_j = \sum_p \chi_s^\sigma$. *The vertex of* M_j *has order* p^{a-e+1}.

(ii) $\Psi_j = \sum_p \chi_s^\sigma + \sum_p \chi_t^\sigma$ *where* $t \sim s$, $v_t = e - 1$, *and the number of p-conjugates of* χ_t *divides* $p - 1$. *The vertex of* M_j *has order* p^{a-e}. *If* $\Psi_j = \Psi_k$, *then* $k = j$.

(iii) $\Psi_j = \sum_p \chi_s^\sigma + \Upsilon$ *where* $\Upsilon \neq 0$ *is a character such that, whenever* $\chi_t \subseteq \Upsilon$, *the number of p-conjugates of* χ_t *is divisible by* p, $v_t \leqq e - 2$, *and each p-conjugate of* χ_t *appears in* Υ *with the same multiplicity as* χ_t. *The vertex of* M_j *has order* p^{a-e}. *If* $\Psi_j = \Psi_k$, *then* $k = j$.

(iv) $p = 2$ *and* $\Psi_j = 2\chi_s$. *The vertex of* M_j *has order* p^{a-e}. *If* $\Psi_j = \Psi_k$, *then* $k = j$.

It is quite possible that the above theorem can be improved—we do not have examples for each type listed.

In case χ_s belongs to a block of defect 1 it is clear that cases A, B (iii), and B (iv) do not occur.[16] Corollary B is a consequence of this observation and our previous results.

---

[16] For the theory of blocks of defect 1, see Brauer [1], [2] and [4, pp. 218–219]. See also Dade [1].

REFERENCES

E. Artin, C. J. Nesbitt and R. Thrall

1. *Rings with minimum condition*, Univ. of Michigan Publ. Math., no. 1, Univ. of Michigan Press, Ann Arbor, Mich., 1944. MR **6**, 33.

R. Brauer

1. *Investigations on group characters*, Ann. of Math. (2) **42** (1941), 936–958. MR **3**, 196.

2. *On groups whose order contains a prime number to the first power.* I, Amer. J. Math. **64** (1942), 401–420. MR **4**, 1.

3. *On the arithmetic in a group ring*, Proc. Nat. Acad. Sci. U.S.A. **30** (1944), 109–114. MR **6**, 34.

4. *On blocks of characters of groups of finite order.* II, Proc. Nat. Acad. Sci. U.S.A. **32** (1946), 215–219. MR **8**, 131.

5. *Zur Darstellungstheorie der Gruppen endlicher Ordnung*, Math. Z. **63** (1956), 406–444. MR **17**, 824.

6. *Zur Darstellungstheorie der Gruppen endlicher Ordnung.* II, Math. Z. **72** (1959/60), 25–46. MR **21** #7258.

E. C. Dade

1. *Blocks with cyclic defect groups*, Ann. of Math. (2) **84** (1966), 20–48. MR **34** #251.

W. Feit

1. *Notes on modular representation theory*, Yale University, New Haven, Conn., 1969.

J. A. Green

1. *Blocks of modular representations*, Math. Z. **79** (1962), 100–115. MR **25** #5114.

2. *A transfer theorem for modular representations*, J. Algebra **1** (1964), 73–84. MR **29** #147.

3. *Some remarks on defect groups*, Math. Z. **107** (1968), 133–150. MR **38** #2222.

G. Keller

   1. *Concerning the degrees of irreducible characters*, Math. Z. **107** (1968), 221–224. MR **38** #3362.

H. Nagao

   1. *A proof of Brauer's theorem on generalized decomposition numbers*, Nagoya Math. J. **22** (1963), 73–77. MR **27** #3714.

M. Osima

   1. *Notes on blocks of group characters*, Math. J. Okayama Univ. **4** (1955), 175–188. MR **17**, 1182.

A. Rosenberg

   1. *Blocks and centres of group algebras*, Math. Z. **76** (1961), 209–216. MR **24** #A158.

L. L. Scott

   1. *Uniprimitive groups of degree kp*, Ph.D. Thesis, Yale University, New Haven, Conn., 1968.

H. Wielandt

   1. *Finite permutation groups*, Lectures, University of Tubingen, 1954/55; English transl., Academic Press, New York, 1964. MR **32** #1252.

UNIVERSITY OF CHICAGO

On the Affine Group over a Finite Field

Louis Solomon

Let G be a finite group with (B,N) pair. Iwahori [3], [4] has shown that the double coset algebra defined by the subgroup B of G has a set $S_1, \ldots, S_n$ of distinguished generators and an irreducible character sgn of degree one such that $\text{sgn}(S_i) = -1$ for $i = 1, \ldots, n$. The double centralizer theory sets up a one-to-one correspondence between irreducible characters of the double coset algebra and irreducible characters of G which appear in the permutation character afforded by G/B. In particular, sgn corresponds to the Steinberg character of G.

Our starting point here is a theorem on finite incidence structures which asserts that an analogue of sgn exists in more general situations. In case the incidence structure is the building [6] defined by a group with (B,N) pair, the character we define is equal to Iwahori's.

Let E be a finite set which is a union of disjoint subsets $E_1, \ldots, E_n$. Suppose E is endowed with a symmetric relation, called incidence, such that two elements of E_i are incident if and only if they are equal. Let F be the set of complete flags of E. Thus F consists of sets $x = \{x_1, \ldots, x_n\}$ where $x_i \in E_i$ and any two elements of x are incident. Let $\Gamma_i(x)$ be the set of complete flags $y = \{y_1, \ldots, y_n\}$ such that $y \neq x$ and $y_j = x_j$ for $j \neq i$. Let M be the vector space over the rational field Q which has F as basis. Define $\gamma_i \in \text{End}(M)$ by $\gamma_i x = \sum_{y \in \Gamma_i(x)} y$. Let C be the algebra of endomorphisms of M generated by $\gamma_1, \ldots, \gamma_n$.

THEOREM 1. *Assume $|\Gamma_i(x)| = q_i$ is independent of x and that $\sum_{i=1}^n (1 + q_i)^{-1} < 1$. Then C has an irreducible character of degree one, which we call* sgn, *such that* $\text{sgn}(\gamma_i) = -1$ *for $i = 1, \ldots, n$.*

Let G be a group of automorphisms of E. Then G acts as a group of linear

---

AMS 1970 *subject classifications*. Primary 50D05, 20C15; Secondary 20G40.

transformations of M and hence defines an algebra A of endomorphisms of M. The centralizer algebra $C(A)$ of A in $\text{End}(M)$ includes C. If $C(A) = C$ then the double centralizer theory shows that G has a rational irreducible character χ corresponding to sgn which may be viewed as an analogue for G of the Steinberg character of a finite group with (B,N) pair. Thus it might be interesting to know which incidence structures E admit a group G of automorphisms for which $C(A) = C$ and to study the irreducible character χ. Henceforth, in this paper, we assume that E is affine geometry of dimension n over the field F_q of q elements.

Let V be a vector space of dimension $n + 1$ over F_q with basis $v_0, v_1, \ldots, v_n$. Let V_∞ be the subspace spanned by $v_1, \ldots, v_n$. Let E_i be the set of i-dimensional subspaces of V not included in V_∞. Let incidence be inclusion. Then $E = E_1 \cup \ldots \cup E_n$ is affine geometry of dimension n. The affine group G consists of all $g \in GL(V)$ such that $gV_\infty \subseteq V_\infty$ and $gv_0 \equiv v_0 \mod V_\infty$. The hypotheses of Theorem 1 are satisfied for sufficiently large q (depending on n) but the conclusion is in fact true for all q.

THEOREM 2. *Let E be affine geometry of dimension n over F_q and let G be the affine group. Then $\gamma_1, \ldots \gamma_n$ generate the centralizer algebra $C(A)$. The irreducible character χ has degree $(q^n - 1)(q^{n-1} - 1) \ldots (q - 1)$.*

We sketch the argument. The main effort is to construct a decomposition of G into double cosets analogous to the Bruhat decomposition for groups with (B,N) pair. Let V_i be the subspace of V spanned by $v_0, \ldots, v_{i-1}$. Then $\{V_1, \ldots, V_n\}$ is a complete flag of E. Let B be its stabilizer in G. The group B affords an idempotent ε of $Q[G]$. Since G is transitive on complete flags, the centralizer algebra $C(A)$ is anti-isomorphic to the double coset algebra $\varepsilon Q[G]\varepsilon$. Under this anti-isomorphism, γ_i corresponds to $\beta_i = |B|^{-1}\sum_{g \in Br_iB} g$ where the elements $r_i \in G$ are defined by

$$r_1v_0 = v_0 + v_1; \quad r_1v_1 = -v_1; \quad r_1v_j = v_j \text{ for } j \neq 0,1$$

and for $i = 2, \ldots, n$

$$r_iv_{i-1} = v_i; \quad r_iv_i = v_{i-1}; \quad r_iv_j = v_j \text{ for } j \neq i - 1, i.$$

Note that $W = \langle r_2, \ldots, r_n \rangle$ is the Weyl group of $GL(V_\infty)$ and that B is a Borel subgroup of $GL(V_\infty)$. Let $s_i = r_ir_{i-1}\ldots r_1\ldots r_{i-1}r_i$. Then $S = \langle s_1, \ldots, s_n \rangle$ is an elementary abelian 2-group and W normalizes S. If $w \in W$ let $l(w)$ be the length of w as a word in $r_2, \ldots, r_n$. If $s \in S$ let $m(s)$ be the length of s as a word in $s_1, \ldots, s_n$.

LEMMA. *Let $w \in W$, $s \in S$, and suppose $i = 2, \ldots, n$. Then*
(i) *If $l(r_iw) > l(w)$ and $m(s_is) > m(s)$, then $r_iBsw \subseteq Br_iswB$.*
(ii) *If $l(r_iw) > l(w)$ and $m(s_is) < m(s)$, then $r_iBsw \subseteq Br_iswB \cup Bs_ir_iswB$.*
(iii) *If $l(r_iw) < l(w)$ and $m(s_{i-1}s) > m(s)$, then $r_iBsw \subseteq Br_iswB \cup BswB$.*
(iv) *If $l(r_iw) < l(w)$ and $m(s_{i-1}s) < m(s)$, then $r_iBsw \subseteq Br_iswB \cup BswB \cup Bs_iswB$.*

Note that precisely one of the cases (i)–(iv) must occur. One should compare this lemma with the corresponding assertion for groups with (B,N) pair, where there is no analogue of the group S and one simply has $r_iBw \subseteq Br_iwB$ if $l(r_iw) > l(w)$ and

$r_i Bw \subseteq BwB \cup Br_i wB$ if $l(r_i w) < (w)$. One proves the lemma, as in the case of Chevalley groups, by using certain commutation relations involving elements of W and S, together with toral and unipotent elements of $\mathrm{GL}(V_\infty)$ and the translations in G. There is a similar lemma in case $i = 1$. Now, since $G = \langle B, r_1, \ldots, r_n \rangle$ one concludes easily that $G = BSWB$. This is analogous to the Bruhat decomposition, although the situation here is more complicated since distinct pairs (s,w) for $s \in S$, $w \in W$ need not define distinct double cosets $BswB$. The final step in the argument is to prove, using $G = BSWB$, that $\beta_1, \ldots, \beta_n$ generate $\varepsilon Q[G]\varepsilon$. Here too there are complications which do not appear in the (B,N) case.

To compute the degree of χ we argue as in [5]. We define a simplicial complex K analogous to Tits's building and show, using results of Folkman [1] on the homology of lattices, that the degree of χ may be computed in terms of the Euler characteristic of K and hence in terms of the Möbius function of the lattice of subspaces of affine geometry. The same character appears in some work of S. I. Gel'fand [2], who defines it as the restriction to G of a certain irreducible character of $\mathrm{GL}(V)$ and proves that it is induced by a linear character of a p-Sylow group of G, where p is the prime divisor of q.

REFERENCES

1. J. Folkman, *The homology groups of a lattice*, J. Math. Mech. **15** (1966), 631–636. MR **32** #5557.

2. S. I. Gel'fand, *Analytic representations of the full linear group over a finite field*, Dokl. Akad. Nauk SSSR **182** (1968), 251–254 = Soviet Math. Dokl. **9** (1968), 1121–1125. MR **38** #1188.

3. N. Iwahori, *On the structure of a Hecke ring of a Chevalley group over a finite field*, J. Fac. Sci. Univ. Tokyo Sect. I **10** (1964), 215–236. MR **29** #2307.

4. N. Iwahori, *On some properties of groups with BN-pairs*, Proc. Sympos. Theory of Finite Groups (Harvard University, Cambridge, Mass., 1968) Benjamin, New York, 1969, pp. 203–212

5. L. Solomon, *The Steinberg character of a finite group with BN-pair*, Proc. Sympos. Theory of Finite Groups (Harvard University, Cambridge, Mass., 1968) Benjamin, New York, 1969, pp. 213–221. MR **40** #220.

6. J. Tits, *Buildings of spherical type* (to appear).

UNIVERSITY OF WISCONSIN

Generalization of Green's Polynomials

T. A. Springer[1]

1. In his work on the complex characters of the finite groups $GL_n(F_q)$, J. A. Green introduced certain polynomials which he showed to be crucial in the description of the irreducible characters. In the present note we shall outline a method to obtain a partial generalization in the case of a reductive algebraic group.

1.1. Let k be a finite field with q elements, of characteristic p. We denote by G a connected reductive linear algebraic group, defined over k. $G(k)$ denotes the finite group of k-rational points of G.

Let $\mathfrak{g}$ be the Lie algebra of G and $\mathfrak{g}(k)$ the finite Lie algebra of its k-rational points.

If S is a finite set, we denote by $|S|$ its cardinal and by $C(S)$ the space of complex valued functions on S. The standard measure on S is the measure for which each point has measure 1.

The method which we follow uses the Fourier transform on the finite abelian group $\mathfrak{g}(k)$. It was inspired by a result of Harish-Chandra. We have to make the following hypothesis on G.

1.2 *There exists a faithful rational representation ρ of G defined over k, such that the symmetric bilinear form $F(\ ,\ )$ on $\mathfrak{g} \times \mathfrak{g}$ defined by $F(X,Y) = Tr(d\rho(X) \cdot d\rho(Y))$ is nondegenerate.*

($d\rho$ denotes the differential of ρ.) It is known that if G is semisimple, such a ρ exists for a group isogeneous to G if G has no simple factors of type A_r and if the characteristic p is good (see [1, E-18, 5.3.]; for good and bad characteristics see loc.cit. E-12). Also if $G = GL_n$, the standard representation ρ satisfies 1.2.

1.3. Let P be a parabolic k-subgroup of G, and let U be its unipotent radical. Let $\mathfrak{p}(k), \mathfrak{u}(k)$ denote the subalgebras of $\mathfrak{g}(k)$ defined by P and U, respectively. $\mathfrak{p}(k)$ is called a cuspidal subalgebra of $\mathfrak{g}(k)$ and $\mathfrak{u}(k)$ its nilpotent radical.

AMS 1970 *subject classifications*. Primary 22E50, 20G05.

[1] The author was partially supported by the Netherlands Organization for the Advancement of Pure Research (Z.W.O.).

For $f \in C(\mathfrak{g}(k))$ put

$$f_P(X) = \int_{\mathfrak{u}(k)} f(X + Y)\, dY$$

where dY is standard measure. f is called a *cusp form* on $\mathfrak{g}(k)$ if $f_P = 0$ for all $P \neq G$. This definition is similar to that of a cusp form on G (see [1, C-5]).

1.4. An element $X \in \mathfrak{g}$ is called *smoothly regular* if the dimension of its Lie algebra centralizer $\mathfrak{z}_\mathfrak{g}(X)$ equals the rank r of G. In that case the group centralizer

$$Z_G(X) = \{g \in G | Ad(g)X = X\}$$

also has dimension r (but the converse need not be true).

If X is semisimple and smoothly regular, the identity component of its centralizer $Z_G(X)$ is a maximal torus. If the characteristic p is not a torsion prime for G, then $Z_G(X)$ is connected (see [1, E-35, 3.19]).

An element of $\mathfrak{g}(k)$ is called *elliptic* if it is contained in the Lie algebra of a minisotropic k-torus of G ([1, E-25, 1.11]).

Let χ be a fixed nontrivial homomorphism of the additive group of k onto C^*. For $X, Y \in \mathfrak{g}(k)$ put

$$\langle X, Y \rangle = \chi(F(X,Y)),$$

with F as in 1.2.

1.5. PROPOSITION. *Let A be a smoothly regular elliptic element of $\mathfrak{g}(k)$. Then the function $X \mapsto \langle X, A \rangle$ is a cusp form on $\mathfrak{g}(k)$.*

The proof is easy. 1.5 is an analogue of a result of Harish-Chandra over $\mathbf{R}$ (this is the result alluded to in 1.1).

To apply 1.5, one has to know when smoothly regular elliptic elements exist. By a theorem of Chevalley, this is so if G is an adjoint semisimple group ([4], the element constructed there is indeed elliptic).

Henceforth, for simplicity we assume either that G is absolutely simple, not of type A_r, and that p is good; or that $G = GL_n$. In the first case Chevalley's theorem shows that smoothly regular elliptic elements exist, and in the second case this is also true, as one easily checks.

1.6 We define a function on $\mathfrak{g}(k) \times \mathfrak{g}(k)$ by

$$\varphi(X, Y; G) = q^{-N} |Z_{G(k)}(Y)|^{-1} \int_{G(k)} \langle X, Ad(g), Y \rangle\, dg.$$

Here $2N$ denotes the number of roots of G, $Z_{G(k)}(Y)$ the centralizer of Y in $G(k)$, and dg is standard measure. If A is as in 1.5, $X \mapsto (X, A; G)$ is a cusp form on $\mathfrak{g}(k)$ which is $G(k)$-invariant.

Let $V(G)$ denote the variety of unipotent elements of G, let $\mathfrak{V}(G)$ denote the variety of nilpotent element of $\mathfrak{g}$. Under the assumption on G made above there is a bijective k-morphism $f: V(G) \to \mathfrak{V}(G)$, commuting with the action of G (see [**1**, E-63, 3.12]). Let $V(G,k)$ be the set of unipotent elements of $G(k)$.

If A is an element of $\mathfrak{g}(k)$, define a function $Q(\ ,A;G)$ on $V(G,k)$ by

$$Q(x,A;G) = \varphi(f(x),A;G).$$

1.7. PROPOSITION. (i) *If* $x \in V(G,k)$, $y \in G(k)$, *then* $Q(yxy^{-1},A;G) = Q(x,A;G)$;

(ii) $q^N Q(x,A;G)$ *is a rational integer*;

(iii) *Suppose A is a smoothly regular elliptic element of $\mathfrak{g}(k)$ and U the unipotent radical of a k-parabolic subgroup of G. Then we have, dx denoting standard measure,*

$$\int_{U(k)} Q(x,A;G)\, dx = 0.$$

Let P be a parabolic k-subgroup of G, and let $P = M \cdot U$ be a Levi decomposition of P. Let h be a complex function defined on the unipotent classes of $M(k)$. Extend h to the unipotent classes of $P(k)$ by defining $h(m \cdot u) = h(m)$. Then the usual formula

$$H(x) = |P(k)|^{-1} \sum_{\substack{y \in G(k); \\ yxy^{-1} \in P(k)}} h(yxy^{-1})$$

defines a function H on $V(G,k)$ which is $G(k)$-invariant. We call H the function on $V(G,k)$ induced by h. With this notation we have

1.8. PROPOSITION. *Let A be a smoothly regular semisimple element of $M(k)$. Then $Q(\ ,A;G)$ is the function on $V(G,k)$ induced by $Q(\ ,A;M)$.*

Finally we have the following orthogonality relation.

1.9. PROPOSITION. *Let $P_i = M_i U_i$ ($i = 1,2$) be two parabolic k-subgroups of G such that the cuspidal subgroups $P_1(k)$ and $P_2(k)$ of $G(k)$ are not associated. Let A_i be a smoothly regular elliptic element in M_i. Suppose there are cusp forms φ_i in $C(M_i)$ such that $\varphi_i(x) = Q(x,A_i;M_i)$ if $x \in V(M_i,k)$ ($i = 1,2$). Then, dx denoting standard measure,*

$$\int_{V(G,k)} Q(x,A_1;G)\, Q(x,A_2;G)\, dx = 0.$$

For the notion of associated cuspidal subgroups see [**1**, C-3].

2. Special cases.

2.1. Assume that $G = GL_n$. The conjugacy classes of unipotent elements of $G(k)$ are parametrized by partitions λ of n (see [**1**, D-5]); let x_λ be a representative of the class parametrized by λ. Also, the classes of maximal k-tori of G are parametrized by such partitions [loc.cit. D-6, D-7]; let T_λ be a representative. Let A_λ be a smoothly regular element of the Lie algebra $\mathfrak{t}_\lambda(k)$ defined by T_λ. There is a function $q_0(n)$ such that A_λ exists if $g \geqq q_0(n)$.

If λ and μ are partitions of n, $Q(\lambda,\mu)$ denotes the integer defined by Green, which is a polynomial function of q (see [1, D-20]).

2.2 THEOREM. *For $q \geq q_0(n)$ we have*
$$Q(\lambda,\mu) = (-1)^{n-1} Q(x_\lambda, A_\mu; GL_n).$$

First let μ be the partition $\{n\}$ with one part. Then A_μ is a smoothly regular elliptic element of $\mathfrak{g}(k)$. By 1.7 (iii), $Q(x_\lambda, A_{\{n\}}; GL_n)$ satisfies Equation (5.4) of [1, D-13], whose solution is unique up to a scalar factor. By [loc.cit. D-13, 5.3 and D-20, 5.16 (v)] we then have
$$Q(x_\lambda, A_{\{n\}}; GL_n) = cQ(\lambda, \{n\}).$$

Taking $\lambda = \{1^n\}$, the partition with n parts (in which case $x_\lambda = e$) we find $c = (-1)^{n-1}$.

In the present case, 1.8 turns out to be property (ii) of [1, D-20, 5.16]. It follows from loc.cit. that $Q(\lambda,\mu)$ is completely determined by $Q(\lambda, \{m\})$ for $m \leq n$ and this property. The assertion then follows.

2.3. 2.2 shows that the $Q(x,A;G)$ generalize indeed Green's $Q(\lambda,\mu)$. The property 1.9 is a partial generalization of the orthogonality relations of [1, D-20, 5.16 (iv)] and can be used to prove these relations in a somewhat simpler way than was done in loc.cit. The assumption made in 1.9 about the existence of φ_i is related to the conjecture of [1, C-21]. In fact, one can conjecture that the values of the character χ_ϕ of loc.cit. on $V(G,k)$ are given by

(2.4) $$\chi_\phi(x) = Q(x,A;G),$$

where A is a suitable smoothly regular element in the Lie algebra of the minisotropic torus involved in the description of χ_ϕ. (2.4) would imply the conjectured formula for the degree of χ_ϕ [1, C-21]. The following result also fits in with the conjectures of loc.cit.

2.5. PROPOSITION. *Assume that p does not divide the order of the Weyl group of G. Let A be a smoothly regular elliptic element of $G(k)$, and let x be a regular unipotent element of $G(k)$. Then*
$$Q(x,A;G) = (-1)^s,$$
where s is the semisimple rank of G.

For the proof one needs the normal form of smoothly regular elements of $\mathfrak{g}$ which was given by Kostant in characteristic 0 [3, p. 381, Theorem 7]. The extension to characteristic p requires the further restriction on p (perhaps unnecessary) which is made in 2.5.

The properties of Green's $Q(\lambda,\mu)$ suggest a number of questions about $Q(x,A;G)$. We mention only one: does the property of $Q(\lambda,\mu)$ of being a polynomial in q generalize to $Q(x,A;G)$ in some way?

References

1. A. Borel et al., *Seminar on algebraic groups and related finite groups*, Lecture Notes in Math., No. 131, Springer-Verlag, Berlin, 1970.

2. J. A. Green, *The characters of the finite general linear groups*, Trans. Amer. Math. Soc. **80** (1955), 402–447. MR **17**, 345.

3. B. Kostant, *Lie group representations on polynomial rings*, Amer. J. Math. **85** (1963), 327–404. MR **28** #1252.

4. J.-P. Serre, "Existence d'éléments reguliers sur les corps finis" in: M. Demazure and A. Grothendieck, *Schémas en groupes*. II, Lecture Notes in Math., no. 152, Springer-Verlag, Berlin, 1970, pp. 342–348.

MATHEMATISCH INSTITUUT DER RIJKSUNIVERSITEIT UTRECHT

A Splitting Principle in Algebraic *K*-Theory

Richard G. Swan

J. Burroughs [3], [4] has proved a splitting principle for the K_0-theory of commutative rings and schemes. I will give here a more abstract form of Burroughs' proof which can be applied to the K_0-theory of groups, Lie algebras, etc.

Let $\mathscr{C}$ be an additive category with a bilinear functor $\otimes : \mathscr{C} \times \mathscr{C} \to \mathscr{C}$. We assume that $\otimes$ is commutative, is associative, and has a unit $\mathcal{O}$. In other words, we are given natural isomorphisms $c : A \otimes B \approx B \otimes A$, $a : A \otimes (B \otimes C) \approx (A \otimes B) \otimes C$, and $u : \mathcal{O} \otimes A \approx A$. Assume further that $\otimes$ has an adjoint $\hom : \mathscr{C}^\circ \times \mathscr{C} \to \mathscr{C}$, so that $\mathrm{Hom}_\mathscr{C}(A \otimes B, C) \approx \mathrm{Hom}_\mathscr{C}(A, \hom(B,C))$. We also assume that we are given (nonadditive) functors $\Lambda^n : \mathscr{C} \to \mathscr{C}$ with natural maps $\otimes^n A \to \Lambda^n A$. The Λ^n are to play the role of exterior powers.

We now wish to define $K_0(\mathscr{C})$ and give it the structure of a λ-ring [1], using $\otimes$ and Λ. To do this, it is necessary to define exactness in $\mathscr{C}$ and to make a number of assumptions concerning the various functors and natural isomorphisms.

We first define a morphism of a system $(\mathscr{C}, \otimes, \hom, \Lambda, \mathcal{O}, c, a, u, \ldots)$ into another one $(\mathscr{C}', \otimes', \hom', \ldots)$ to be a functor $F : \mathscr{C} \to \mathscr{C}'$ together with natural isomorphisms $F(A \otimes B) \approx FA \otimes FB$, $F\mathcal{O} \approx \mathcal{O}'$, $F(\hom(A,B)) \approx \hom(FA,FB)$, $F\Lambda^n(A) \approx \Lambda^n F(A)$ such that these are compatible with the various natural isomorphisms. In other words, the diagrams

$$\begin{array}{ccc} F(A \otimes B) \approx FA \otimes FB & \quad & \mathrm{Hom}(A, \hom(B,C)) \approx \mathrm{Hom}(A \otimes B, C) \\ \downarrow \quad\quad\quad \downarrow & & \downarrow \quad\quad\quad\quad\quad\quad \downarrow \\ F(B \otimes A) \approx FB \otimes FA & & \mathrm{Hom}(FA, \hom(FB,FC)) \approx \mathrm{Hom}(FA \otimes FB, FC) \end{array}$$

etc. are required to commute.

If R is a commutative ring, let $\mathscr{M}_R$ be the category of R-modules and $\mathscr{P}_R$ the full

---

AMS 1970 *subject classifications.* Primary 18F30, 16A54; Secondary 18F25, 13D15.

subcategory of finitely generated projective R-modules. We define $\otimes$, hom, etc. on these categories in the usual way.

We now assume that there is given a morphism $T: \mathscr{C} \to \mathscr{P}_R$. We will say that a sequence $A \to B \to C$ in $\mathscr{C}$ is exact if the sequence $TA \to TB \to TC$ is exact in $\mathscr{M}_R$. The functor T is required to have the following properties:

(a) T is faithful.

(b) If $A \xrightarrow{i} B \xrightarrow{j} C \to 0$ is exact, then $C = \operatorname{ckr} i$.

(c) If $0 \to A \xrightarrow{i} B \xrightarrow{j} C \to 0$ is exact, then $A = \ker j$.

(d) If $B \xrightarrow{j} C \to 0$ is exact, there is an exact sequence $0 \to A \to B \xrightarrow{j} C \to 0$.

(e) If $f: A \to B$ and $\operatorname{ckr} T(f)$ (computed in $\mathscr{M}_R$) lies in $\mathscr{P}_R$, then there is an exact sequence $A \to B \xrightarrow{j} C \to 0$.

EXAMPLE. If $\mathscr{C}$ is a full subcategory of an abelian category $\mathscr{A}$ and T extends to a faithful exact functor $T: \mathscr{A} \to \mathscr{M}_R$, then all these properties are obviously satisfied.

For technical reasons, we will also assume that the ring R is noetherian. The general case will be discussed in more detail elsewhere.

We now define $K_0(\mathscr{C})$ in the usual way. It is the abelian group with generators $[A]$ for all objects A in $\mathscr{C}$, with relations $[A] = [A'] + [A'']$ for all exact sequences $0 \to A' \to A \to A'' \to 0$ in $\mathscr{C}$. Since tensoring preserves exactness, $K_0(\mathscr{C})$ is a commutative, associative ring with unit where $[A][B] = [A \otimes B]$.

If $A \in \mathscr{C}$, a (finite) filtration of A is defined to be a sequence of morphisms $0 = F_{-1}A \to F_0A \to \ldots \to F_nA = A$ which, under T, becomes isomorphic to a filtration of the R-module $T(A)$. We say that the filtration is good if there are exact sequences $0 \to F_{i-1}A \to F_iA \to B_i \to 0$. If so, $[A] = \sum [B_i]$ in $K_0(\mathscr{C})$. We write $\operatorname{gr}_i A = B_i$.

If $0 \to A' \to A \to A'' \to 0$ is exact, the Koszul filtration of $\Lambda(A)$ is defined by letting $F_i \Lambda^n(A)$ be the image of $\Lambda^{n-i}(A') \otimes \Lambda^i(A)$ in $\Lambda^n(A)$. This exists by our assumptions on T (im = ker ckr). Using T, we easily see that $\operatorname{gr} \Lambda(A) \approx \Lambda(A') \otimes \Lambda(A'')$. Let $\lambda_t(A) = \sum [\Lambda^n(A)] t^n \in K_0(\mathscr{C})[[t]]$. Then $\lambda_t(A) = \lambda_t(A')\lambda_t(A'')$ so $K_0(\mathscr{C})$ is a λ-ring [1]. Our objective is the following theorem:

THEOREM 1. $K_0(\mathscr{C})$ is a special λ-ring [1].[1]

COROLLARY 2. The Adams operations defined by $\sum \psi^k(x) t^k = -d/dt (\log \lambda_{-t}(x))$ satisfy the relations $\psi^k(xy) = \psi^k(x)\psi^k(y)$, $\psi^h \psi^k(x) = \psi^{hk}(x)$, and $\psi^p(x) \equiv x^p \pmod{p}$ for prime p.

The following are some cases to which this theorem applies.

EXAMPLE 1. $\mathscr{C} = \mathscr{P}_R$ for a commutative noetherian ring R. More generally, let $\mathscr{C}$ be the category of locally free sheaves on a noetherian scheme X. This case is treated by Burroughs in [3],[4]. Let $X = \bigcup U_i$ where the U_i are open and affine. Then $\amalg U_i$ is affine. Let R be its ring. The map $\amalg U_i \to X$ induces T.

EXAMPLE 2. Let G be a group and R a commutative noetherian ring. Let $\mathscr{C}$ be the category of RG-modules which are finitely generated and projective as R-modules.

---

[1] I would like to thank H. Bass for bringing this paper to my attention.

Define $\otimes$, Λ by taking these over R and letting G act in the usual way. Let T forget the G-action.

EXAMPLE 3. Let R be a commutative noetherian ring and $\mathfrak{G}$ a Lie algebra over R. Let $\mathscr{C}$ be the category of R-modules with $\mathfrak{G}$-action which are finitely generated and projective as R-modules. Define $\otimes$, Λ, T as in Example 2.

We can generalize Examples 2 and 3 by considering any Hopf algebra over R which is coassociative, associative, and cocommutative.

In all these examples, T extends to a faithful exact functor on an abelian category. Here is an example where it does not.

EXAMPLE 4. Let R be a commutative noetherian ring, let Alg_R be the category of commutative associative R-algebras with unit, and let G be a covariant functor from Alg_R to groups (e.g. a group scheme over R). A matrix representation of G is a natural transformation $G \to GL_n$. This suggests the following definition: A G-module is an R-module M together with a natural action of $G(A)$ on $A \otimes_R M$ for each $A \in \text{Alg}_R$. Naturality means that if $A \to B$ in Alg_R then $A \otimes_R M \to B \otimes_R M$ is semilinear with respect to $G(A) \to G(B)$. Let $\mathscr{C}$ be the category of G-modules which are finitely generated and projective over R. Define $\otimes$, Λ as in Example 2. Let T forget the G-action. Here T does not extend to an exact functor on an abelian category. Kernels misbehave since $A \otimes_R -$ is not exact. However, our assumptions on T still hold.

We prove Theorem 1 by establishing a splitting principle. A morphism of $T: C \to \mathscr{P}_R$ to $T': \mathscr{C}' \to \mathscr{P}_{R'}$ means a morphism $F: \mathscr{C} \to \mathscr{C}'$ together with a ring homomorphism $\varphi: R \to R'$ such that the diagram

$$\begin{array}{ccc} \mathscr{C} & \to & \mathscr{C}' \\ \downarrow & F & \downarrow \\ \mathscr{P}_R & \underset{\Phi}{\to} & \mathscr{P}_{R'} \end{array}$$

commutes (with $\Phi(P) = R' \otimes_R P$).

If $A \in \mathscr{C}$, define $\text{rk } A = \max \text{rk } T(A)_\mathfrak{p}$ over all prime ideals $\mathfrak{p}$ in R. Clearly $\dim A = \text{rk } A$ in the terminology of [1].

THEOREM 3. *Let $A_1, \ldots, A_r \in \mathscr{C}$. Then there is a morphism $F: (\mathscr{C}, T, R) \to (\mathscr{C}', T', R')$ such that*:

(1) *Each $F(A_i)$ has a good filtration with quotients of rank 1.*
(2) *$K_0 F: K_0(\mathscr{C}) \to K_0(\mathscr{C}')$ is injective.*

This implies Theorem 1 as in [1]. To prove Theorem 3, it will obviously suffice to consider a single A and to produce $F: \mathscr{C} \to \mathscr{C}'$ such that there is an exact sequence $0 \to B \to F(A) \to C \to 0$ where the theorem applies to B and C, e.g. $\text{rk } B \leq 1$ or $\text{rk } B < \text{rk } A$, etc. We consider first the case where $\text{rk}_\mathfrak{p} T(A) = n$ is constant. Let $S = S(A)$ be the symmetric algebra on A. This is the graded algebra whose component $S^k(A)$ is defined to be the coequalizer of $\otimes^k A \rightrightarrows \otimes^k A$ using all permutations. We consider graded S-modules, i.e., graded objects $P = (P_n)$ in $\mathscr{C}$ with a map $S \otimes P \to P$ satisfying the usual conditions. Such a P defines a graded module

$T(P) = (T(P_n))$ over $T(S) = S(TA)$ and therefore a quasi-coherent sheaf[2] $\tilde{P}$ on the projective scheme X over R defined by $S(TA)$. We consider only those P such that P is generated by the P_i for $i \leq$ some i_0, $P_i = 0$ for $i \ll 0$, and such that $\tilde{P}$ is locally free. These will be the objects of $\mathscr{C}'$. A morphism $P \to P'$ in $\mathscr{C}'$ will be a "stable morphism," i.e. a collection $f_n: P_n \to P'_n$ for $n \geq$ some n_0 giving an S-module map $c_{n_0}(P) \to P'$ where $c_{n_0}(P)_n = P_n$ for $n \geq n_0$ and is 0 for $n < n_0$. We identify $f = (f_n)$ with $g = (g_n)$ if $f_n = g_n$ for $n \geq$ some N. Let $\mathscr{L}$ be the category of locally free sheaves on X. We define $\mathscr{C}' \to \mathscr{L}$ by $P \to \tilde{P}$, and $\mathscr{L} \to \mathscr{P}_{R'}$ as in Example 1. Let $T': \mathscr{C}' \to \mathscr{P}_{R'}$ be the composition of these functors. The hypothesis that R is noetherian is needed to verify the required properties of T'. For example, if $f: P \to Q$ and if $T'(f)$ is an isomorphism, then $\ker T(f)$ and $\operatorname{ckr} T(f)$ define zero sheaves. Since these modules are finitely generated, they must be 0 in large dimensions. Thus $T(f)$ is a stable isomorphism and so is f itself. In particular, $P_n \to \Gamma(X, P(n))$ is an isomorphism for large n. In proving (d) and (e) we use the fact [3] that if L is a locally free sheaf, then $\Gamma(X, L(n))$ lies in $\mathscr{P}_R$ for large n.

Define $(P \otimes Q)_n = \coprod P_i \otimes Q_j$ over $i + j = n$ and let $P \otimes' Q = P \otimes_S Q$ be the coequalizer of $P \otimes S \overline{\otimes} Q \to P \otimes Q$. Define $\Lambda^n(A)$ as the pushout of

$$\otimes^n(A) \to \otimes^n_S(A)$$
$$\downarrow$$
$$\Lambda^n(A)$$

If $P \in \mathscr{C}$, then $P \otimes S$ is an object of $\mathscr{C}'$. Set $\hom'(P \otimes S, Q) = \hom(P, Q)$ for $Q \in \mathscr{C}'$. If $M \in \mathscr{C}'$, there is an exact sequence $F' \to F \to M \to 0$ where F and F' are finite direct sums of objects of the form $P \otimes S$. Define $\hom'(M, Q)$ to be the kernel of $\hom'(F, Q) \to \hom'(F', Q)$. This exists by the properties of T and gives an adjoint to $\otimes$. Define $F: \mathscr{C} \to \mathscr{C}'$ by $P \mapsto P \otimes S$.

For any n define $S(n)_i = S_{n+i}$. Since $S_1 = A$, the map $S_i \otimes A = S_i \otimes S_1 \to S_{i+1} = S(1)_i$ defines a map $F(A) = S \otimes A \to S(1)$, and we have an exact sequence $0 \to B \to F(A) \to S(1) \to 0$ with $\operatorname{rk} B < \operatorname{rk} A$.

To show $K_0(F)$ injective, we define, as in [3], a map

$$\theta: K_0(\mathscr{C}') \to K_0(\mathscr{C})[[t]]$$
$$K_0(\mathscr{C})[t] = H$$

by $\theta(P) = \sum [P_n] t^n$. If $\sigma_t(A) = \sum [S_n(A)] t^n$, we have $\theta(F(P)) = \theta(P \otimes S) = [P]\sigma_t(A)$. Therefore $\theta(F(x)) = \sigma_t(A)x$ for $x \in K_0(\mathscr{C})$. If $F(x) = 0$, then $\sigma_t(A)x = 0$ in H so $\sigma_t(A)x = f(t)$ is a polynomial in t. Now just as in [2], [3] we have $\lambda_t(A)\sigma_t(A) = 1$ so $x = \lambda_t(A)f(t)$. But $\lambda_t(A) = 1 + \ldots + [\Lambda^n(A)]t^n$. Let $C = \Lambda^n(A)$. Since $\operatorname{rk}_\mathfrak{p} TA = n$ for all $\mathfrak{p}$, we have $\operatorname{rk}_\mathfrak{p} C = 1$ for all $\mathfrak{p}$. By using T we see that $C \otimes \hom(C, \mathcal{O}) \to \mathcal{O}$ is an isomorphism so $[C]$ is a unit in $K_0(\mathscr{C})$. Since $\lambda_t(A)f(t) = x$ which involves no t, we must have $f(t) = 0$.

---

[2] The use of sheaves and schemes is easily avoided here, but it enables us to use the results of [3] without repeating the calculations.

Finally, we consider the case where $\text{rk}_p TA$ is not constant. Let $n = \text{rk } A$, and let $C = \Lambda^n A$. This time we get an exact sequence $0 \to C \otimes \text{hom}(C,\mathcal{O}) \to \mathcal{O} \to J \to 0$. Write $I = C \otimes \text{hom}(C,\mathcal{O})$. By applying $A \otimes -$ we get $0 \to A \otimes I \to A \to A \otimes J \to 0$. Using T we see that $\text{rk } A \otimes J < n$. Now $D = (A \otimes I) \oplus J^n$ has constant rank n. Construct $\mathscr{C}'$ using $S(D)$ as above. In $\mathscr{C}'$ we have $0 \to E \to F(D) \to S(1) \to 0$. Now $I \otimes J = 0$ and $0 \to I \to \mathcal{O} \to J \to 0$ so $I \otimes I \cong \mathcal{O} \otimes I = I$. Therefore $D \otimes I \approx A \otimes I$. Since $I \otimes F(D) \approx F(I \otimes D)$ we get $0 \to E \otimes I \to F(A \otimes I) \to S(1) \otimes I \to 0$ with $\text{rk } E \otimes I < n$ and $\text{rk }(S(1) \otimes I) \leq 1$.

ADDED IN PROOF. A. Grothendieck recently informed me that a theorem similar to the one proved here has already appeared in the Berthellot-Illusie-Grothendieck Seminar SGA6 VI 3.3. The theorem is proved for ringed topi and covers all the examples given here. I have not yet seen these notes and so cannot comment on the relation between the two proofs. Further results in this direction have been obtained by Deligne.

REFERENCES

1. M. F. Atiyah and D. O. Tall, *Group representations, λ-rings and the J-homomorphism*, Topology **8** (1969), 253–297. MR **39** # 5702.
2. H. Bass, *Algebraic K-theory*, Benjamin, New York, 1968.
3. J. Burroughs, *The splitting principle in algebraic K-theory*, Topology **9** (1970), 129–135.
4. ——— Thesis, University of Chicago, Chicago, Ill., 1968.

UNIVERSITY OF CHICAGO

Classification of Simple Groups of Order $p \cdot 3^a \cdot 2^b$, p a Prime

David Wales[1]

Thompson's N-group paper [7] infers that if a simple group order is divisible by only three distinct primes, the primes are 2,3 and one of 5,7,13,17. The known such simple groups all have orders $p \cdot 3^a \cdot 2^b$ where p is one of these primes. In this report we discuss the proof that the only simple groups of order $p \cdot 3^a \cdot 2^b$ are the known groups A_5, A_6, $PSp_4(3)$, $PSL_2(7)$, $PSL_2(8)$, $U_3(3)$, $PSL_2(13)$, $PSL_2(17)$.

The first case, $p = 5$, is treated by Brauer in [2]. The techniques are mainly those of modular character theory. Particularly important are the results of [1]. A general lemma applicable to the other primes is as follows. In a simple group of order $p \cdot 3^a \cdot 2^b$, a p-element is self centralizing. This is used by Brauer for the modular character theory for the prime $p = 5$.

The cases $p = 7, 17, 13$, are treated in [9], [10], [11]. In these papers, the results and techniques of the N-group paper as well as the modular character theory are used. In particular it is assumed that a minimal counterexample contains a nonsolvable local subgroup. In many cases, a maximal nonsolvable 2- or 3-local subgroup must contain a full Sylow 2- or 3-group. When this can be forced, modular character theory yields contradictions. When it cannot be forced, useful statements about the action on the 2- or 3-group can be inferred. In these cases, the characterizations of $E_2(3)$ and $Sp_4(3)$ occurring in the N-group paper are employed. In the case $p = 7$ it is necessary to show that $2 \sim 3$ and $3 \in \pi_4$. In the case $p = 13$, the results do not hold as the 3-local subgroups may be nonsolvable. This case is handled using the techniques of [7].

Let G be a simple group of order $p \cdot 3^a \cdot 2^b$. By the result of Brauer [2], a p-element is self centralizing. This means that the centralizers of all 2- or 3-elements are $\{2,3\}$-groups and so solvable. This means that all 2- or 3-local subgroups are 2- or 3-

---

AMS 1970 *subject classifications.* Primary 20D05; Secondary 20C20.

[1] Research was partially sponsored by GP 13626.

constrained. Using the terminology of the N-group paper [7], $2 \in \pi_2, \pi_3$, or π_4. That is, a Sylow 2-group P of G has no normal abelian group of rank 3, P has a normal abelian group of rank 3 and P normalizes a group of odd order, or P has a normal abelian group of rank 3 and P normalizes no group of odd order.

We discuss first the results of [8]. The case $2 \in \pi_2$ is treated by Janko and Thompson [6]. That is, if $2 \in \pi_2$, G is listed in the paper [6]. Otherwise, we can assume $2 \in \pi_3$ or π_4. The case $2 \in \pi_3$ is handled using the A-signalizer functor theory of Gorenstein and Walter. We really need only a special case applicable when $2 \in \pi_3$ and the 2-local subgroups are 2-constrained. This is discussed in [5, Part IV]. Using this result, the Thompson transitivity theorem [3, Theorem 8.5.4], and some easy lemmas, we show that $2 \notin \pi_3$ [8, Lemma 6.1]. This means finally $2 \in \pi_4$.

We now link up the existence of nontrivial modular blocks with group theoretic properties. In particular it is shown that if the q-local subgroups are q-constrained, q a prime, and $D \neq e$ is the defect group for a nonprincipal q-block, then $O_{q'}(N(D)) \neq e$. Furthermore, $O_{q'}(N(d)) \neq e$ for any $d \neq e$ in D. Conversely, if $O_{q'}(N(D)) \neq e$, then D is contained in the defect group for some nonprincipal q-block. We now use this result together with the fact $2 \in \pi_4$ to ensure that any nonprincipal 2-block is of defect zero. These techniques yield Theorem 1 of [4]. Now some character theory results special to the case $|G| = p \cdot 3^a \cdot 2^b$ show the same is true of 3. Furthermore $3 \notin \pi_3$. It follows there is a character of degree 3^a and one of degree 2^b [8, Lemma 6.2].

It is now evident that the characterizations of $E_2(3)$ and $Sp_4(3)$ in [7, §§8, 9] may be applied. Specifically, if the 3-local subgroups can be shown to be solvable, we must only show $2 \sim 3$ and $3 \in \pi_4$. That is, we must show the Sylow 3-group has an abelian subgroup of rank 3 and there is a $\{2,3\}$ group with specific properties described in detail in [7, §2]. It is shown in [9, Lemma 8.4] that if $b \geq 14$ this is related to showing there is a subgroup of G isomorphic to $z_2 \times z_3 \times z_3$ or $z_2 \times z_2 \times z_3$.

We discuss briefly the case $p = 7$ [9]. It is shown first that a maximal 3-local nonsolvable subgroup M must contain a full Sylow 3-group. This is done by considering the irreducible actions of $PSL_2(7)$, $U_3(3)$ or $PSL_2(8)$ on a characteristic 3-vector space. We need consider only actions in which a 7-element acts fixed point freely. These actions are very limited and in fact impossible for $PSL_2(8)$. From these it follows that if $M/L \cong PSL_2(7)$ or $U_3(3)$, L a 3-group, $J(L) = J(P)$ where P is a Sylow 3-group of M. It follows that a maximal 3-local subgroup M contains a full Sylow 3-group.

We now eliminate this case by restricting the characters χ_1 and χ_2 of degrees 2^b and 3^a to M. In the case that a Sylow 7-normalizer has order 42, the hardest case, the principal 7-block contains 7 characters whose degrees are all congruent to ± 1 (mod 7). The degrees are 1, 2^b, 3^a plus four more. If $L = O_3(M)$, $|L| = 3^{6s}$ as a 7-element acts fixed point freely on L. As $3^a \equiv \pm 1 \pmod{7}$, M/L must have a Sylow 3-group whose order is congruent to ± 1 (mod 7). This means $M/L \cong U_3(3)$ or $G_2(2)$. A counting of the Sylow 3-groups eliminates $G_2(2)$ and a delicate bounding of orders of induced characters from M eliminates $U_3(3)$. These arguments can be carried out ostensibly because of the following simple result: $1/7 \leq 2^b/3^a \leq 7$. This follows because $2^{2b} + 3^{2a} \leq 7 \cdot 3^a \cdot 2^b$ as there are irreducible characters of degree 2^b and 3^a.

We can now assume the 3-local subgroups are solvable and so there must be a nonsolvable 2-local subgroup by [7]. This case is treated by character theory somewhat as above if a maximal 2-local nonsolvable subgroup contains a full Sylow 2-group. It is really shown only that $2 \sim 3$ and $3 \in \pi_4$. If a maximal nonsolvable 2-local subgroup M does not contain a full Sylow 2-group, we derive properties of $O_2(M)$ and try to show $2 \sim 3$ and $3 \in \pi_4$. A difficult case appears to be a maximal subgroup isomorphic to either a split or nonsplit extension of $z_2 \times z_2 \times z_2$ by $PSL_2(7)$ acting naturally as $SL_3(2)$. At any rate, using character theory and this subgroup we finally force $2 \sim 3$ and $3 \in \pi_4$ and are done.

The case $17 \cdot 3^a \cdot 2^b$, [10], is similar to $7 \cdot 3^a \cdot 2^b$. The group theoretical analysis is actually easier because there is only one group $PSL_2(17)$ to worry about and it is not a characteristic 2- or characteristic 3-group. For example, it is easy to force maximal 2- or 3-local subgroups to contain a full Sylow 2- or 3-group. However, the character theory is not so easy.

In the case $13 \cdot 3^a \cdot 2^b$, [11], it is fairly easy to force a Sylow 13-normalizer to have order $13 \cdot 12$ and to force the 2-local subgroups to be solvable. This forces some 3-local subgroup to be nonsolvable. By working with this configuration and using the ideas of [7, §7), a contradiction is reached.

BIBLIOGRAPHY

1. R. Brauer, *On groups whose order contains a prime number to the first power*. I, Amer. J. Math. **64** (1942), 401–420. MR **4**, 1.

2. ———, *On simple groups of order* $5 \cdot 3^a \cdot 2^b$, Bull. Amer. Math. Soc. **74** (1968), 900–903. MR **38** #4552.

3. D. Gorenstein, *Finite groups*, Harper & Row, New York, 1968. MR **38** #229.

4. ———, *On finite groups of characteristic 2 type*, Inst. Haute Études Sci. Publ. Math. No. 36 (1969), 5–13.

5. ———, *Centralizers of involutions in finite simple groups*, Lectures, Oxford Group Theory Conference, 1969.

6. Z. Janko and J. G. Thompson, *On finite simple groups whose Sylow 2-subgroups have no normal elementary subgroups of order 8*, Math. Z. **113** (1970), 383–397.

7. J. G. Thompson, *Nonsolvable finite groups all of whose local subgroups are solvable*, Bull. Amer. Math. Soc. **74** (1968), 383–437; Pacific J. Math. **33** (1970), 451–537 (balance to appear).

8. D. Wales, *Simple groups of order* $p \cdot 3^a \cdot 2^b$ J. Algebra **16** (1970), 183–190.

9. ———, *Simple groups of order* $7 \cdot 3^a \cdot 2^b$ (to appear).

10. ———, *Simple groups of order* $17 \cdot 3^a \cdot 2^b$ (to appear).

11. ———, *Simple groups of order* $13 \cdot 3^a \cdot 2^b$, (to appear).

CALIFORNIA INSTITUTE OF TECHNOLOGY

Direct Summands in Representation Algebras

W. D. Wallis

1. **Introduction.** The results of this paper appear in [5]. I have taken this opportunity to reorganize the argument along a line suggested in conversation by Dr. S. B. Conlon, to whom I express my gratitude. Proofs identical to those in [5] have been omitted.

We will assume that G is a finite group and F a complete local Noetherian ring whose residue class field has characteristic $p \,(\neq 0)$. We write $\{1\}$ for the one-element group. $A(FG)$ is the usual representation algebra of isomorphism classes of FG-representation modules, and is linear over the complex field $\mathfrak{C}$; $A_H(FG)$ is the subalgebra spanned by the classes of modules with vertex at most H. The structure of $A(FG)$ is given by the structures of the $A_H(FG)$, where H is a p-subgroup of G (see, for example, [1]). We need consider only the case where H is normal in G. (For reasons see [5, p. 395].)

We shall derive a direct sum decomposition of $A_H(FG)$. In [5] we worked for most of the time in $A_H(FG)$; here we shall do most of our algebra in $A(F(G/H))$, and apply the results to $A(FG)$ at the end.

We have discovered that our decomposition is closely related to the decomposition discovered by Conlon in [2]. This relationship will be the subject of another paper [3].

We will denote the isomorphism-class of the representation modules $\mathscr{M}$, $\mathscr{M}^X$, $\mathscr{M}^x$ by M, M^X, M^x, and so on. When $Y \leq X$ define a map

$$\theta = \theta(Y,X): A(FY) \to A(FX)$$

by $M \mapsto M^X$ and by assuming $\mathfrak{C}$-linearity.

AMS 1970 *subject classifications.* Primary 20C20.

2. **Decomposition of the projective ideal.** In this section X is any group; $Y, Z, \ldots$ are subgroups of X. We write $P(X)$ for the projective ideal of $A(FX)$, spanned by the classes of FX-representation modules with vertex $\{1\}$, and define

$$P^X(Y) = P(Y)\theta(Y,X).$$

Note that $P^X(Y) = P^Z(Y)\theta(Z,X)$ when $Y \leq Z$, and that $P^X(Y)$ is spanned by the M^X where $\mathcal{M}$ is an indecomposable FY-representation module with vertex $\{1\}$.

The following results correspond to Lemmas 13 and 15 of [5]. Note that the proofs are much simpler.

LEMMA 1. *If Y and T are any subgroups of X then*

$$P^X(Y)P^X(T) \subseteq \sum_{x \in X} P^X(Y^x \cap T).$$

PROOF. Suppose $\mathcal{Q}$ is an FY-module and $\mathcal{R}$ is an FT-module. Then, from Mackey's "tensor product" theorem [4, p. 325],

$$(1) \qquad \mathcal{Q}^X \otimes \mathcal{R}^X \cong \bigoplus_x ((\mathcal{Q}^x)_{Y^x \cap T} \otimes \mathcal{R}_{Y^x \cap T})^X;$$

where x runs through a complete set of (Y,T)-double coset representatives in X. In particular, if $\mathcal{Q}$ and $\mathcal{R}$ both have vertex $\{1\}$ then each tensor product on the right has vertex $\{1\}$. Hence the class of the module on the right of (1) lies in $\sum P^X(Y^x \cap T)$; since classes like Q^X and R^X respectively span $P^X(Y)$ and $P^X(T)$, the result follows.

LEMMA 2. *If Y is X-conjugate to a subgroup of Z, then $P^X(Y) \subseteq P^X(Z)$.*

PROOF. Let Y and Y^x be conjugate subgroups of X. Then the indecomposable FY^x-modules with vertex $\{1\}$ are precisely the modules $\mathcal{M}^x$ where $\mathcal{M}$ is an indecomposable FY-module with vertex $\{1\}$. So the $(M^x)^X$ span $P^X(Y^x)$. Since $(\mathcal{M}^x)^X \cong \mathcal{M}^X$, this means that $P^X(Y^x)$ is spanned by a subset of $P^X(Y)$; the converse holds similarly, so $P^X(Y^x) = P^X(Y)$, and we can assume $Y \leq Z$.

If Q is any indecomposable FY-representation module with vertex $\{1\}$, then Q^Z has vertex $\{1\}$, so $Q^Z \in P(Z)$. But Q can be any basis element of $P(Y)$. So

$$(2) \qquad P^Z(Y) \subseteq P(Z),$$

$$P^X(Y) = P^Z(Y)\theta(Z,X) \subseteq P(Z)\theta(Z,X) = P^X(Z).$$

As a corollary to (2), $P^X(Y)$ is a subalgebra of $P(X)$ for any Y. But we can show [5, Theorem 4] that $P^X(Y)$ is an ideal of $A(FX)$. So it is an ideal of $P(X)$. $P(X)$ is a finite direct sum of copies of $\mathfrak{C}$, so $P^X(Y)$ also has this form. Therefore it has an identity element; call it J_Y.

From Lemma 2, $J_Y \in P^X(Z)$, and J_Y is idempotent. So $J_Z = J_Y + (J_Z - J_Y)$ is an orthogonal idempotent decomposition; so

$(3) \qquad P(X)J_Y$ *is an ideal direct summand of* $P(X)J_Z$ *when* $Y \leq Z$.

Lemma 1 tells us $J_Y J_T \in \sum P^X(Y^x \cap T)$, so

(4)[1] $$\{P(X)J_Y\}\{P(X)J_T\} \subseteq \sum_{x \in X} P(X)J_{(Y^x \cap T)}.$$

Define $\pi(Z)$ to be a complete set of non-X-conjugate subgroups of Z, and $\pi'(Z)$ to be $\pi(Z)\setminus\{Z\}$, and write $Q(Z) = \sum P(Y)J_Y$, where $\sum$ means algebra sum over $Y \in \pi'(Z)$. By (3), $Q(Z)$ is a finite sum of direct summands of $P(X)J_Z$, so $Q(Z)|P(Z)J_Z$; consequently there is an algebra $R(Z)$, defined by $P(X)J_Z = Q(Z) \oplus R(Z)$. As $P(X)J_Z$ has an identity element, so does $R(Z)$; call the identity element K_Z. We can prove

THEOREM 1. $P(Z)J_Z = \bigoplus R(Y)$, where $\bigoplus$ is algebra direct sum over $Y \in \pi(Z)$.

(The proof is formally identical to that of Theorem 21 of [5].) This decomposition corresponds to the orthogonal idempotent decomposition

(5) $$J_Z = \sum K_Y.$$

3. **Decomposition of $A_H(FG)$.** We now identify X above with G/H; $Z, \ldots$ are $R/H, \ldots$ for groups satisfying $H \leqq Z \leqq G. \ldots$

We define an algebra monomorphism $\psi = \psi_R : A(F(R/H)) \to A(FR)$ as follows: $\mathcal{M}\psi$ is an FR-module whose elements $m\psi$ correspond to the elements m of the $F(R/H)$-module $\mathcal{M}$, and if $r \in R$ then $r(m\psi) = ((rH)m)\psi$. $M\psi$ is the class of $\mathcal{M}\psi$; ψ is extended to $A(F(R/H))$ by linearity.

In our notation the identity element of $P(G/H)$ is $J_{G/H}$. Theorem 3.17(a) of [1] says that $J_{G/H}\psi_G$ is the identity element of $A_H(FG)$. So using (5) we obtain

THEOREM 2. $$A_H(FG) = \bigoplus A_H(FG)K_Y\psi_S$$

where $\bigoplus$ is algebra direct sum over all Y in $\pi(G/H)$, and S is defined by $S/H = Y$.

We can now continue to compare Theorem 2 with Conlon's decomposition in [1], obtaining the refinement in (27) of [5], and to evaluate $A_H(FG)K_Y\psi_S$ in various cases. For this the reader is referred to [5].

REFERENCES

1. S. B. Conlon, *Relative components of representations*, J. Algebra **8** (1968), 478–501. MR **36** #6475.
2. ———, *Decompositions induced from the Burnside algebra*, J. Algebra **10** (1968), 102–122. MR **38** #5945.
3. S. B. Conlon and W. D. Wallis, *The identity of certain representation algebra decompositions*, Bull. Austral. Math. Soc. **3** (1970) 73–74.
4. C. W. Curtis and I. Reiner, *Representation theory of finite groups and associative algebras*, Pure and Appl. Math., vol. 11, Wiley, New York, 1962. MR **26** #2519.
5. W. D. Wallis, *Decomposition of representation algebras*, J. Austral. Math. Soc. **10** (1969), 395–402.

UNIVERSITY OF NEWCASTLE, AUSTRALIA

---

[1]This equation is not used in this paper. However, it is needed in the proof of Theorem 1, so I have included it for completeness.

Representations of Chevalley Groups in Characteristic p

W. J. Wong[1]

If G_K is a Chevalley group over a field K of prime characteristic p, the irreducible representations of G_K over K form a natural object of study. The basic results have been obtained by Steinberg[6], who showed that if K is perfect, then each irreducible rational representation of G_K over K is a tensor product of representations obtained from certain basic representations by composing them with field automorphisms. These basic representations were obtained by "integrating" the irreducible restricted representations of a restricted Lie algebra associated with the group, which had been studied earlier by Curtis [2]. The present author had obtained the main results previously for the groups $SL(n,K)$, $Sp(2n,K)$ by different means, involving reduction (mod p) from the characteristic 0 case [7]. We now extend this method to the other types of groups, in the hope that some additional insight may be gained. Possibly the most interesting result obtained is a simple necessary and sufficient condition under which an irreducible module in the characteristic 0 case remains irreducible (mod p).

We begin with a complex semisimple Lie algebra $\mathfrak{g}$ of rank n, with Cartan decomposition

$$\mathfrak{g} = \mathfrak{h} \oplus \sum_{r \in S} \mathfrak{g}_r.$$

We take a Chevalley basis consisting of the fundamental coweights $H_1, \ldots, H_n$ in $\mathfrak{h}$ and root vectors X_r, $r \in S$. Let $\mathfrak{U}_z$ be the subring of the universal enveloping algebra $\mathfrak{U}$ of $\mathfrak{g}$ generated by the elements $X_r^m/m!$, $r \in S$, $m \geq 0$. An *admissible lattice* on a (finite-dimensional) $\mathfrak{g}$-module V is an additive subgroup V_z of V which is invariant under $\mathfrak{U}_z$ and which has a $\mathbf{Z}$-basis that is a vector space basis of V. Every

---

AMS 1970 *subject classifications*. Primary 20G05, 20G40; Secondary 20C20.

[1] Research partially supported by National Science Foundation grant GP-11342 at the University of Notre Dame.

g-module has admissible lattices (Kostant [3]). If K is any field and G_K the group of rational points over K of the simply connected Chevalley group of type g (e.g., $G_K = SL(n + 1, K)$ when $\mathfrak{g} = A_n$), then the K-vector space $\bar{V} = V_z \otimes_z K$ becomes a G_K-module. For a given V, a change of admissible lattice V_z may give a different G_K-module $\bar{V}$, but the irreducible constituents of $\bar{V}$ are not changed. Our approach is to study the structure of such G_K-modules $\bar{V}$.

Suppose that K is finite, with q elements. (The essential aspects of the situation appear already in the finite case.) For each dominant integral function μ on $\mathfrak{h}$, let $V(\mu)$ be the irreducible g-module of highest weight μ. We denote by P_q the set of dominant integral functions μ such that $0 \leq \mu(H_i) \leq q - 1$, for all i. By studying Brauer characters, we classify the irreducible G_K-modules as follows.

THEOREM 1. *For each $\mu \in P_q$, there is an irreducible G_K-module $F(\mu)$ over K, such that the irreducible constituents of $\bar{V}(\mu)$ are $F(\mu)$, occurring with multiplicity 1, and possibly certain $F(\nu)$ with ν lower than μ. The $F(\mu)$ form a complete set of absolutely irreducible G_K-modules.*

The modules $\bar{V}(\mu)$ are not always irreducible. To see when they are, we introduce a new idea. It is easily seen that g has an automorphism θ such that $\theta(X_r) = -X_{-r}$, for all roots r. From this fact it follows that, if V is any g-module, there exists a nondegenerate bilinear form (,) on V such that

$$(vX_r, w) = (v, wX_{-r})$$

for all $v, w \in V$, $r \in S$. We call such a form *contravariant*. If V is irreducible and v_0 is a vector of highest weight, the contravariant form on V is unique to within scalar multiplication, and is completely determined if we require that $(v_0, v_0) = 1$. Then $V_z = v_0 \mathfrak{U}_z$ is an admissible lattice on V, and the form is integral and symmetric on V_z. Let $d(V)$ denote the discriminant of this integral form. We then have the following irreducibility criterion.

THEOREM 2. *If $\mu \in P_q$, then $\bar{V}(\mu)$ is irreducible if and only if $d(V(\mu)) \not\equiv 0 \pmod{p}$.*

In any given case, the criterion can be computed, although not as easily as Springer's sufficient condition [5].

In order to describe the irreducible modules $F(\mu)$ in general, we need to assume that g satisfies a certain condition. Let $\lambda_1, \ldots, \lambda_n$ be the fundamental weights, $\lambda_i(H_j) = \delta_{ij}$. Then we consider the following condition.

(∗) For $i = 1, \ldots, n$, there exists a g-module V_i with λ_i as highest weight, occurring with multiplicity 1, such that V_i has an admissible lattice $(V_i)_z$ and a contravariant form which is integral and unimodular on $(V_i)_z$.

This condition has been verified when g is simple of type $A_n, B_n, C_n, D_n, E_6, F_4, G_2$, and it seems likely that it is always satisfied.

If $\mu \in P_q$, let $m_i = \mu(H_i)$, and form the m_ith symmetric powers $V_i^{(m_i)}, (V_i)_z^{(m_i)}$ of $V_i, (V_i)_z$. Then

$$V = V_1^{(m_i)}, \otimes \cdots \otimes V_n^{(m_n)}$$

has μ as highest weight, with multiplicity 1, and has

$$V_z = (V_1)_z^{(m_1)} \otimes \cdots \otimes (V_n)_z^{(m_n)}$$

as admissible lattice. Take a vector v_0 of weight μ such that $\mathbf{Z}v_0$ is a direct summand of V_z as an Abelian group. Form the G_K-module $\bar{V}$ and let X be the submodule of $\bar{V}$ generated by $\bar{v}_0 = v_0 \otimes 1$. Then, under the assumption (∗), we can prove that X has a unique maximal submodule Y, and $X/Y \cong F(\mu)$. Furthermore, Y has an explicit description as the intersection of X with the kernels of certain linear functionals on $\bar{V}$. Also, we can prove the tensor product theorem of [5], expressing the $F(\mu)$ in terms of those for which $0 \leq \mu(H_i) \leq p - 1$, for all i.

We remark finally that essentially these methods and results were used by Brauer and Nesbitt [1] and by Mark [4] to find the irreducible Brauer characters of $SL(2,q)$ and $SL(3,q)$.

References

1. R. Brauer and C. Nesbitt, *On the modular characters of groups*, Ann. of Math. (2) **42** (1941), 556–590. MR **2**, 309.

2. C. W. Curtis, *Representations of Lie algebras of classical type with applications to linear groups*, J. Math. Mech. **9** (1960), 307–326. MR **22** #1634.

3. B. Kostant, *Groups over* Z, Proc. Sympos. Pure Math., vol. 9, Amer. Math. Soc., Providence, R.I., 1966, pp. 90–98. MR **34** #7528.

4. C. Mark, *The irreducible modular representations of the group GL*(3p), Thesis, University of Toronto, Ontario, Canada, 1938.

5. T. A. Springer, *Weyl's character formula for algebraic groups*, Invent. Math. **5** (1968), 85–105. MR **37** #2763.

6. R. Steinberg, *Representations of algebraic groups*, Nagoya Math. J. **22** (1963), 33–56. MR **27** #5870.

7. W. J. Wong, *On the irreducible modular representations of finite classical groups*, Thesis, Harvard University, Cambridge, Mass., 1959.

UNIVERSITY OF NOTRE DAME

Nilpotent Elements in Representation Rings

Janice Rose Zemanek

Let G be a finite group and let k be a field of characteristic p, $p \neq 0$. The integral representation ring $a(kG)$ is the abelian group whose generators are isomorphism classes $[M]$ of kG-modules, with addition $[M] + [M'] = [M \oplus M']$. Multiplication is given by the tensor product, $[M][M'] = [M \otimes_k M']$.

We shall consider the question as to when does $a(kG)$ contain nonzero nilpotent elements. Green [2], [3] and O'Reilly [4] showed that $a(kG)$ is semisimple if $p \nmid |G|$ or if the p-Sylow subgroups of G are cyclic. Conlon [1] and Wallis [5] have proved that $a(kG)$ is semisimple if $p = 2$ and the p-Sylow subgroups of G are of type (2,2).

Our main result is in the opposite direction.

THEOREM. *Let p be an odd prime. Suppose that G has a noncyclic p-Sylow subgroup. Then $a(kG)$ contains a nonzero nilpotent element.*

The proof is an easy consequence of the following result.

THEOREM. *Let p be an arbitrary prime. Let Z_p^* be the ring of p-adic integers. Let G be a group with a noncyclic p-Sylow subgroup. Then $a(Z_p^*G)$ contains a nonzero nilpotent element* [6].

In the case where p is odd, we then show that the nilpotent element defined in $a(Z_p^*G)$ is not in the kernel of the ring homomorphism

$$a(Z_p^*G) \to a(\bar{Z}G)$$

given by $[M] \to [M/pM]$. Since $\bar{Z} = Z/pZ$ is a subfield of k, the monomorphism

$$a(\bar{Z}G) \to a(kG)$$

defined by $[M] \to [k \otimes_{\bar{Z}} M]$ gives a nonzero nilpotent element in $a(kG)$.

AMS 1969 subject classifications. Primary 1640; Secondary 1644.

References

1. S. B. Conlon, *The modular representation algebra of groups with Sylow 2-subgroup $Z_2 \times Z_2$*, J. Austral. Math. Soc. **6** (1966), 76–88. MR **34** #250.

2. J. A. Green, *The modular representation algebra of a finite group*, Illinois J. Math. **6** (1962), 607–619. MR **25** #5106.

3. ———, *A transfer theorem for modular representations*, J. of Algebra **1** (1964), 73–84. MR **29** #147.

4. M. F. O'Reilly, *On the semisimplicity of the modular representation algebra of a finite group*, Illinois J. Math. **9** (1965), 261–276. MR **30** #4841.

5. W. D. Wallis, *Factor ideals of some representation algebras*, J. Austral. Math. Soc. **9** (1969), 109–123. MR **39** #5723.

6. J. Zemanek, *On the semisimplicity of integral representation rings*, Ph.D. Thesis, University of Illinois, Urbana, Ill., 1970.

LOUISIANA STATE UNIVERSITY

SUBJECT INDEX

$AG(d, q_2)$, 68
$\alpha_\chi = \max\{|G|_p/|B| \mid B \in \mathcal{V}_\chi\}$, 19
Abelian normal subgroup, 37
Adams operations, 156
Adjoints to restriction, 42
Admissible lattice, 169
    Irreducibility criterion, 170
Admissible transformations, 113
Affine group, 146
Algebra, 111
    Bounded type, 111
    Finite type, 111
    Indecomposable representations, 111
    Matrix questions, 113
    Polynomial part, 120
    Strongly unbounded type, 111
    Unbounded type, 111
Algebraic maps, 39, 40, 42, 43
    Transfer, 42, 44
Alternating form, 73
Alternating group, 95
Automorphism, 25

(B, N) pair, 91
Basic set, 8
Bass-order, 130, 137–142
    Completely primary, 133
Bass-ring, 131, 137
Block
    Type of block, 10
Bounded type, 111
Brauer characters, 124
Brauer homomorphism, 140
Brauer's first fundamental theorem 141
Brauer's second fundamental theorem, 142
Brauer's second main theorem on blocks, 124

$\mathscr{C}$-decomposition, 29
    Isomorphic refinements, 30
Cancellation, 85
Cancellation for modules, 31
Category of G-functors, 42
Characteristic powers, 1
Characterization of characters, 124

Character ring functor, 58
Characters, 49, 51
    Brauer characters, 124
    Character table, 97
    Characterization of characters, 124
    Extension of character, 51
    Invariant, 51
    Irreducible, 3, 80–82, 146
    Orthogonality relations, 49
    Permutation, 73
    Principal indecomposable, 123
    q-modular (Brauer), 97
Character table, 97
Chevalley groups, 1
    Finite, 13
    Hecke algebra, 2
    Irreducible representation, 13, 169
Cohomology ring functor, 59
Completely primary Bass-order, 133
Compounds, 92
1-1 condition, 141
Conjugacy classes, 51, 78
Contravariant form, 170
Conway group, 108
Cusp form, 150

Δ-theorem, 117
Decomposition matrix, 124
Decomposition numbers, 138
Defect 0 and 1, 142
Defect base, 60
Defect-basis, 44
Defect group, 61, 139
Doubly transitive group, 67

Exchange property, 30
Exponentials, 40
Extension of character, 51

Factorizable groups, 77
F. C. subgroup, 117
Fields of characteristic p, 99
    Finite Chevalley groups, 13

Finite groups
   $AG(d, q_2)$, 68
   Adjoints to restriction, 42
   Algebraic maps, 39, 40, 42, 43
   Alternating group, 95
   Automorphism, 25
   (B, N) pair, 91
   Basic set, 8
   Brauer homomorphism, 140
   Chevalley groups, 1
   1-1 condition, 141
   Conjugacy classes, 51, 78
   Conway group, 108
   Doubly transitive group, 67
   Exponentials, 40
   Fixed point subgroup, 25
   G-functor, 42, 43, 44, 57, 124
   $GL(n, C)$, 37
   Generalized quaternion groups, 74
   Hecke algebra, 91
   Hyperelementary groups, 97
   $\mathcal{M}(\phi)$, 20
   Indecomposable representations, 89
   Index parameters of G, 91
   Integral representation ring, 173
   Leech lattice, 108
   Linear group, 37
   Metacyclic groups, 65, 79
   Nonsimplicity, 47
   Onto condition, 141
   $PSU_4(3)$, 107
   Permutation character, 73
   Projective oG-modules, 88
   Real representations, 66
   Reflection representation, 92
   Relative Grothendieck rings, 44, 99
   Representation modules, 165, 166
   Representations of finite groups, 99
   Suzuki group, 107
   Sylow 2-subgroups of simple groups, 53
   Units in $\Omega(G)$, 41
   X-graded Clifford system, 20
Finite type, 111
Fixed point subgroup, 25
Frobenius algebra, 49
   Characters, 49
Frobenius–Schur formula, 98
Fundamental module, 89

G-algebra, 59

G-functor, 42, 43, 44, 57, 124
   Category of G-functors, 42
   Character ring functor, 58
   Cohomology ring functor, 59
   Defect base, 60
   Defect group, 61, 139
   G-algebra, 59
   Grothendieck ring functors, 59
   Subgroup category, 58
   Transfer theorem, 61
$GL(n, C)$, 37
Genera of R-lattices, 85
Generalized polynomial identity, 119
Generalized quaterion groups, 74
Generic degree, 2
Generic ring, 1, 92
Genus, 85
   Restricted genus, 87
Gorenstein-ring, 131–133, 137, 138
Green's polynomials, 149
Grothendieck ring functors, 59
Group
   Lie, 13
   Primitive, 37
   Simple, 13, 161
Group ring, 117

Hecke algebra, 2, 91
Hyperelementary groups, 97

Indecomposable lattice, 137, 140
Indecomposable representations, 89, 111
   Infinite type, 111
Index parameters of G, 91
Induction theorems, 44, 45
Infinite type, 111
$\mathcal{M}(\phi)$, 20
Integral representation, 85
Integral representation ring, 173
Invariant character, 51
   π-blocks, 124
   Parabolic type, 3
Irreducibility criterion, 170
Irreducible character, 3, 80–82, 146
   Generic degree, 2
   Green's polynomials, 149
Irreducible representation, 13, 169
Isomorphic refinements, 30

SUBJECT INDEX

Krull-Schmidt (-Azumaya) theorem, 29

λ-ring, 155
Lattice
  Cancellation, 85
  Genera of R-lattices, 85
  Genus, 85
  Indecomposable, 137, 140
  Local direct factor of M, 87
Leech lattice, 108
Lie algebra
  Cusp form, 150
Lie groups, 13
Linear group, 37
Local direct factor M, 87

$M_{\mathscr{C}}(\mathscr{C}_1\text{-free})$, 21, 22
$M_{\mathfrak{D}(G)}(M\text{-homogeneous})$, 21, 22
Matrix questions, 113
  Admissible transformations, 113
Metacyclic groups, 65, 79
  Split, 65
Modular theory of permutation representations, 137
Modules
  Admissible lattice, 169
  Brauer's first fundamental theorem, 141
  $\mathscr{C}$-decomposition, 29
  Cancellation for modules, 31
  Contravariant form, 170
  Decomposition numbers, 138
  Defect 0 and 1, 142
  Exchange property, 30
  Fundamental, 89
  Krull-Schmidt (-Azumaya) theorem, 29
  $M_{\mathscr{C}}(\mathscr{C}_1\text{-free})$, 21, 22
  $M_{\mathfrak{D}(G)}(M\text{-homogeneous})$, 21, 22
  Schur index, 97
  Tensor product theorem, 171
  Vertex, 165, 166

Nilpotent radical, 118
Nonsimplicity, 47

o-order, 85
Onto condition, 141
Order
  Bass-order, 130, 137, 142
  Integral representation, 85
Orthogonal idempotent decomposition 166, 167
Orthogonality relations, 49

π-blocks, 124
  Brauer's second main theorem on blocks, 124
$PSU_4(3)$, 107
Parabolic type, 3
  Partially ordered set, 114
Permutation character, 73
Polynomial identity, 118
Polynomial part, 120
Prime rings, 117
Primitive, 37
Principal indecomposable characters, 123
Projective ideal, 166
  Orthogonal idempotent decomposition, 166, 167
Projective oG-modules, 88

q-modular (Brauer) character, 97

Real representations, 66
Reflection representation, 92
  Compounds, 92
Relative Grothendieck rings, 44, 99
  Defect-basis, 44
  Induction theorems, 44, 45
Representation algebra, 165
Representation modules, 65, 66
  Projective ideal, 166
  Representation algebra, 165
Representation of finite groups, 99
  Modular theory of permutation representations, 137
Restricted genus, 87
Ring
  Adams operations, 156
  Bass-ring, 131, 137
  Δ-theorem, 117
  Generalized polynomial identity, 119
  Gorenstein-ring, 131–133, 137, 138
  Group ring, 117
  Integral representation ring, 173
  λ-ring, 155
  o-order, 85
  Polynomial identity, 118
  Prime, 117
  Semiprime, 118
  Semisimple, 120
  Splitting principle, 155

Schur index, 97
  Frobenius-Schur formula, 98

Semiprime rings, 118
  Nilpotent radical, 118
Semisimple rings, 120
  Sets of primes, 123
Simple groups, 13, 162
Split metacyclic groups, 65
Splitting principle, 155
Strongly embedded subgroup, 69
Strongly unbounded type, 111
Subgroup category, 58
Subgroups
  Abelian normal subgroup, 37
  F. C. subgroup, 117
  Strongly embedded, 69
  Weakly closed, 26
Suzuki group, 107
Sylow 2-subgroups of simple groups, 53
System of (B, N)-pairs of type (W, R), 1
  Characteristic powers, 1

Generic ring, 1, 72

Tensor product theorem, 171
Transfer, 42, 44
Transfer theorem, 61
Type of block, 10

Unbounded type, 111
  Strongly, 111
Units in $\Omega(G)$, 41

Vector space
  Alternating form, 73
Vertex, 165, 166
  $\alpha_\chi = \max\{|G|_p/|B| \,|\, B \in \mathcal{V}_\chi\}$, 19

Weakly closed subgroups, 26

X-graded Clifford system, 20